D0214776

# GEOMETRY

## AN EXERCISE IN REASONING

KEN SEYDEL

Skyline College, San Bruno, California

SAUNDERS COLLEGE PUBLISHING    Philadelphia

Saunders College Publishing
West Washington Square
Philadelphia, PA   19105

To My
Parents

Geometry

ISBN   0-7216-8070-4

© 1980 by Saunders College Publishing/Holt, Rinehart and Winston.   Copyright under the Inter-
national Copyright Union.   All rights reserved.   This book is protected by copyright.   No part of it
may be reproduced, stored in a retrieval system, or transmitted in any form or by any means,
electronic, mechanical, photocopying, recording, or otherwise, without written permission from the
publisher.   Made in the United States of America.   Press of W. B. Saunders Company.   Library of
Congress catalog card number 78-64723

56    026    10  9  8  7  6  5  4  3

# TO THE INSTRUCTOR

The text is tailored for the *community college student*, and is an attempt to present the beauty and interest of Geometry without losing the formality. When looking for a suitable text to use in our geometry course we experienced a general dissatisfaction with what was available. In talking with other instructors I discovered few, if any, using a book about which they were enthusiastic. Every book sales representative that visited me was asked what was available in geometry and most said they wished they had a really good book. There are some very good high school books available, but they are too lengthy for a one semester course. One can not use selected sections from such a book owing to the highly sequential nature of the material. The community college student is a unique person—needing material at a more mature level than that of the high school but often not being a sufficiently strong student to be able to handle the more rigorous material intended for the university student. Consequently, I undertook the compilation of my class notes into something that could be called a text. It is designed principally for a three unit course, but has sufficient material and exercises for a five unit course.

The development of a geometry should ideally use all the available tools—transformations, vectors, coordinates, and synthetic. Some of the traditionally difficult theorems become ridiculously simple with a different approach. However, the student does not have all these tools available, and we do not have the time to develop them. Therefore, the approach is much as Euclid did it. Some transformations have been used as a tool, but this is not a Transformational Geometry. Transformation concepts are used mainly as a justification for the type of "superpositioning" used by Euclid.

The main emphasis should be on logic and reasoning. An attempt is made to tie the deductive reasoning discussed in the beginning sections to a "string of conditional statements", which is then related to the standard two column proof. The students' thoughts are guided with an analysis in the more difficult proofs.

There are some sections that stand alone and could be omitted. They are : Non-Euclidean Geometries, Concurrency, Composite Figures, The Pythagorean Theorem As Area, and The Golden Mean. I feel the omission of these sections would be a loss; however, I would suggest these omissions rather than running short of time and not completing the last two sections on surface and volume. They are therefore placed last in the text. The opening page of each chapter indicates the best place to include them if time permits.

Finally, it is suggested that the problems marked with an asterisk (*) not be regularly assigned. For the most part they are difficult or time consuming. Students should be encouraged to look them over and occasionally attempt one that interests them.

The main objective in the teaching of geometry over these past 2200 years or so has been the hope that the student will learn to think deductively. Recently, however, the educational psychologists have demonstrated rather poor "transfer of learning". That is, studying and learning a deductive science does not necessarily make one a deductive thinker. Consequently, I try to be more direct in the teaching of thought patterns. I try to show how a "string of conditionals" will form a proof. It is my hope that the student will be led to see the thought process in the work he or she does and recognize proofs as an example of good deductive reasoning.

A second objective of geometry is, of course, the learning of some geometric relationships. These I try to make more meaningful with applied problems. An attempt is made to develop the main geometric facts needed in further work in mathematics, the sciences, and everyday usage. The development of these relationships from a very few axioms and postulates is similar to the traditional. The mathematics is honest, even in the few instances where some "rigor" has been sacrificed in the interest of pedagogy.

An attempt has been made to keep the presentation informal— almost conversational, and to keep the symbolism to a minimum. An attempt has also been made to keep the book concise—keeping the number of theorems to the minimum necessary for the development of future topics. It has been difficult to resist the temptation to include many beautiful and interesting theorems. Some minor theorems have been included as exercises.

You will find many concepts and exercises never before included in an elementary geometry text. It is hoped that these, along with applied problems and recreational items, will help keep a high interest level. It is also hoped that the majority of the material and exercises can be handled successfully by the average student, and that this success will generate a confidence in, and enjoyment of, mathematics.

Whatever merit this volume might have is largely due to the encouragement and excellent help received from many sources. The initial credit goes to Jack Snyder—Vice President of the College Division, W. B. Saunders Company—who suggested that I write the book. Next, my appreciation to the students in my classes who helped to find intelligible explanations. Also to Paul Goodman and Jim Ertman of Skyline College, who class tested the material and made many helpful suggestions.

Thanks to the American Book Company for their permission to use a number of examples and exercises from an excellent book, now out of print, by Shute, Shirk, and Porter.

Our reviewers were:

> Richard Dawdy, Long Beach City College;
> Robert Denton, Orange Coast Community College;
> Richard De Tar, Sacramento City College;
> David Minor, Chabot College;
> Harold Oxsen, Diablo Valley College;
> Gene Sellers, Sacramento City College;
> Ronald R. Young, Harrisburg Area Community College;
> Philip Gillett, University of Wisconsin.

Both the praise for things done well and the criticism of things done poorly are much appreciated.

Special thanks goes to the staff at Saunders who guided and supported me through the process of publication: Bill Karjane, Mathematics Editor; David Milley, Manuscript Editor; Lorraine Battista, Design Manager; Juliana Kremer, Administrative Assistant of the College Division; and Frank Polizzano, Production Manager for College Titles. Their skills transformed ideas into a book!

# TO THE STUDENT

I have some good news and some bad news for you.

First the good news: Geometry is one of the most useful and interesting subjects a person can study. When you finish this course you will see relationships you previously overlooked, see beauty and patterns where you never saw them before, and be more competent in many practical areas.

And now the bad news: Geometry is not a spectator sport. It is a participant activity. It is something you *do*. You will not learn Geometry by watching your instructor work problems. Therefore it would be wise to plan to spend an hour out of class for every hour in class if you are to get the most benefit from the course. More bad news: Geometry, by its very nature, requires some memorization. In order to justify statements that you make, it is necessary to give reasons. Acceptable reasons are axioms, postulates, definitions, and theorems, which therefore need to be remembered. As an aid they are listed in the appendix.

For the course you will need a metric ruler, a protractor, and a compass. You should also have an understanding of Elementary Algebra.

The problems in the exercises marked with an asterisk (*) are intended only for the student with additional time and interest. I would suggest that you look them over and see if there are some you want to tackle. Sometimes it is worthwhile to work on a problem even if you don't solve it, and if you are successful it will increase your confidence and give you a feeling of competency.

If you enjoy puzzles and mental games you should enjoy this course. Approach it with a positive attitude and the idea that it is an interesting subject. And don't be discouraged when there are problems that you can't do—it is not expected that you be 100% perfect. But when you make a mistake, learn from it and don't make that same mistake again. (Make a different mistake next time.)

If you begin to think more logically, learn some useful geometric relationships, and find that you can appreciate and enjoy mathematics, then—regardless of your grade—this course will have been immensely successful!

Any comments or suggestions will be most appreciated, and I will answer all correspondence.

Ken Seydel
Skyline College
San Bruno, CA 94066

v

# CONTENTS

A copy of the title page of the first English translation of
Euclid's "The Elements". By Sir Henry Billingsley in 1570.
(Courtesy of The British Museum.)

# A BRIEF HISTORY OF GEOMETRY

The meaning of the word geometry (from the Greek) is "earth measure" (geo—earth, metry—measure). This possibly originated from the surveyors, rope stretchers, who surveyed the flood plains of the Nile after each rainy season. I would prefer to take "earth measure" in the larger sense: the study of the measures of all things on the earth.

I suppose the history of geometry really begins with the history of humankind. When we first tried to understand the world around us, we began to develop geometric concepts. "It's as far to the volcano as it is to the river," is a geometrical relationship. So perhaps this discussion should be confined to the history of geometry that is recorded. Geometry is the earliest recorded mathematical science, and had its beginnings with the Egyptians and Babylonians around 3000 BC.

Dating is very difficult. The earliest dated event of human history is the introduction of the Egyptian calendar in 4241 BC. (This calendar had 12 months of 30 days plus 5 feast days, and was more accurate than the Roman calendar that was used in Europe until the introduction of the current calendar in 1582).

Geometry of a high degree of accuracy was used in the construction of the pyramids around 2900 BC. For example, the pyramid of Gizeh used 2 million huge stones weighing up to 54 tons each. They were hauled 600 miles and cut to an accuracy of 1 part in 10,000. Tablets found at Nippur record a Babylonian geometry of 1500 BC. Chinese geometry is recorded as early as 1100 BC. All of the earlier geometries were of a practical nature, with no attempts to generalize or make use of deductive thought. The Egyptians, Babylonians, Chinese, Indians, and Romans all had fairly well developed practical geometry.

The Greeks were the originators of demonstrative geometry; that is, the use of deductive methods to formulate proofs. The Classical Greeks of 660–300 BC were the real originators. The Greek historian Herodotus (5th century BC) conjectured that geometry began with the surveying of the Nile floodplains. Thales (640–546 BC) was a successful, wealthy, well-traveled man of commerce who became interested in geometry. He studied the geometry of the Egyptians and began to expand it and to generalize. One of his pupils, Pythagoras (580?–501?) developed the geometry into an abstract science. Pythagoras also profited from the Egyptian practical geometry. He was advised to visit Egypt by Thales. Much interesting literature is available about the Order of the Pythagoreans—a school (more like a secret society) founded by Pythagoras. Other important Greeks of this period were Hippocrates (470–?)—not the great physician; Plato (429–348); Eudoxus (408–355); Aristotle (382–322)—although known primarily for his systematization of deductive logic, he improved some geometric definitions and discussed continuity; Euclid (330?–275?); Archimedes (287?–212)—very accomplished in geometry and mechanics, he located pi between $3\frac{10}{71}$ and $3\frac{1}{7}$ using polygons of 96 sides; and Apollonius (260–200).

Many other mathematicians followed, but they did not achieve the glory of the early Greeks. Eratosthenes (c. 230 BC), Hipparchus (c. 140 BC), Heron (c. 110 BC), Menelaus (c. 100 BC), Claudius Ptolemy (85?–165? AD), Pappus (c. 340 AD), Theon of Alexandria (c. 390 AD), and his daughter Hypatia (375–415)—the first recorded woman mathematician.

The greatest of these was Euclid. He is reported to be the first Professor of Mathematics at the University of Alexandria. He took a few Postulates and a few Axioms and developed thirteen books of demonstrative geometry. When asked by Ptolemy if there was an easier way to master geometry, he is reported to have replied, "There is no royal road to Geometry." His work, with only minor revisions and corrections, has been taught for over 2000 years. The title of his work was *"The Elements"*. A comment on *The Elements* by Sir Thomas Heath: "This wonderful book, with all its imperfections, which are indeed slight enough when account is taken of the date at which it appears, is, and will doubtless remain, the greatest mathematical textbook of all time. Scarcely any other book except the Bible can have been circulated more widely the world over, or been more edited or studied."

The Dark Ages, beginning with the fall of Alexandria in 641 AD, were also the dark ages for mathematics. Only minor discoveries were made during the next thousand years. Fortunately, the Arabs recognized the value of much of the work in the library at Alexandria and preserved and translated it into Arabic. They maintained this knowledge throughout the Dark Ages, and also adopted and refined the Hindu numeral system. At the end of the Dark Ages, this knowledge began to find its way back into Europe. In the twelfth and thirteenth centuries, this early Greek knowledge was translated from Arabic to Latin and many universities started to relearn the "lost" knowledge. The plagues and the Hundred Years' War slowed things down during the fourteenth century, but things picked up again during the next two centuries. The invention of the printing press was a great boon to the dispersal of knowledge. So in the seventeenth and eighteenth centuries mathematics in general and geometry in particular were ready for further growth and development.

Euclid's *Elements* was first printed in Latin in Venice in 1482. The first English translation was by Billinglsey in London in 1570. (See the illustration at the beginning of this section.) Descartes (1596–1650) unified the algebra with the geometry with his introduction of coordinates and analytic geometry. In 1639 Girard Desargues (1593–1662) introduced projective geometry. Solid analytic geometry was developed by Antoine Parent (1666–1716) and furthered by Alexis Claude Clairaut (1713–1765), one of twenty children of a mathematics teacher (who was quite good at multiplying!).

Over the years mathematicians had been bothered by one of the postulates of the *Elements*—Euclid's parallel postulate. It seemed to them that it ought to be possible to prove it rather than accept it as a postulate. One of the earliest noteworthy attempts in this direction was by a Jesuit priest named Girolamo Saccheri (1667–1733). He tried an indirect proof, but was unable to come up with a contradiction. His work did, however, lay the foundations for the non-Euclidean Geometries. These were geometries that used a different parallel postulate from Euclid's, but were still consistent and sufficient deductive sciences.

Carl Friedrich Gauss (1777–1855) was probably the first to realize that a different set of postulates was possible. He did not publish, however, because it was such a revolutionary concept and he did not want to upset the Church. Janos Bolyai (1802–1860), a Hungarian cavalry officer, developed a "new" geometry in 1824 and published in 1832. Nicolai Ivanovitch Lobachevsky (1793–1856), a professor at the University of Kazan at age 23, published a non-Euclidian geometry in 1826. George Friedrich Bernard Riemann (1826–1866), one of Gauss' best students and a lecturer at the University of Gottingen, also developed a non-Euclidean geometry, and its application to physical phenomena by Einstein has shown that Euclid's geometry is not a necessity of thought, and is not even the most convenient geometry to apply to existing space. Einstein's General Theory of Relativity makes use of Riemannian geometry. Arthur Cayley (1821–1895) unified the Euclidean and non-Euclidean geometries into one comprehensive theory.

Various branches of geometry are still being actively developed even today, and it is now quite difficult to determine just where geometry leaves off and where other branches of mathematics begin.

# GEOMETRY

*One of the main purposes of studying geometry is to de-*
*velop good methods of thought.* Photo courtesy of The
Rodin Museum.

# CHAPTER 1

# REASONING

In this chapter we will look at some of the many reasons for studying geometry—functional, logical, and aesthetic—and try our hand at some game-type problems that will illustrate the need for proper reasoning techniques. We will discuss what it means to "prove" a statement, what is meant by "deductive reasoning", and the form of reasoning called the syllogism. Finally, we will look at some of the more common abuses of good reasoning.

## 1. Why we study geometry

Although every person taking a course in geometry has his or her own motivations, some of the more common reasons are, unfortunately: "Because I have to," and "My counselor said I needed it." What these responses are saying is that geometry is a prerequisite to some sequences of courses that lead to the student's vocational or professional objective. These are rather superficial reasons and should be pursued further. Why is geometry a prerequisite for other courses such as trigonometry and calculus?

To be honest with you, the geometric relationships used in these courses are few in number, and could be presented (but not logically developed) in a week or two. Why has the study of geometry been required of serious students for nearly 2000 years? The answers to these questions yield some of the more meaningful reasons for the study of geometry:

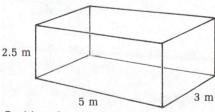

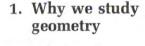

1 liter covers
15 sq. m

*Problem 1*
How much paint is necessary to paint the room?
(neglecting windows and doors)
(Don't paint the floor!)

3

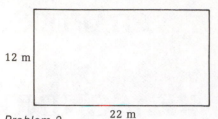

12 m

22 m

Use 1 bag for
each 30 sq. m

*Problem 2*
How much fertilizer does the lawn require?

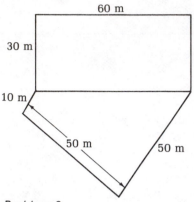

60 m

30 m

10 m

50 m

50 m

*Problem 3*
Are these two lots about the same size?

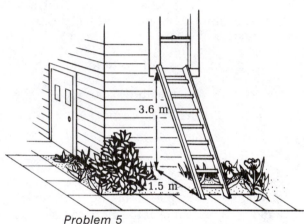

30 cm

20 cm

*Problem 4*
How much gas will this can hold?

3.6 m

1.5 m

*Problem 5*
How tall must the ladder be?

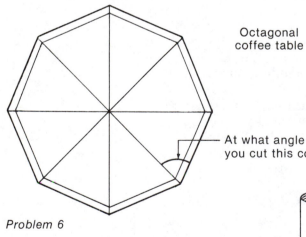

Octagonal
coffee table

At what angle do
you cut this corner?

*Problem 6*

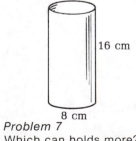

16 cm

8 cm

4 cm

16 cm

*Problem 7*
Which can holds more?
Which can requires the least amount of tin to construct?

4

First, we study geometry for **information,** both useful and aesthetic. It is necessary to have a knowledge of geometric relationships (formulas in most of these examples) in order to be able to answer useful questions such as these. You have no doubt learned many formulas, such as the area of a square, rectangle, or circle; the circumference of a circle; the volume of a box, sphere, or cylinder. In this course we will learn other shapes and their formulas, and we will *prove* that these formulas are always valid.

Hopefully, you will also learn from geometry something about symmetry, proportion, and regular shapes that will help you to better appreciate the geometric beauty, both natural and man-made, in the world around you.

The great book of Nature lies ever open before our eyes and the true philosophy is written in it—but we cannot read it unless we have first learned the language and the characters in which it is written. It is written in mathematical language and the characters are triangles, circles, and other geometrical figures.—*Galileo*

The original reason for geometry in the curriculum, and it is still considered to be the most important reason to teach geometry, is to develop some **patterns of thought.** After two thousand years, demonstrative geometry is still taught because it offers the best vehicle, and the clearest example, for introducing the student to the type of thinking which is necessary both in the sciences and in everyday experiences as well. What we are referring to here is *deductive logic*; but more about that later!

Geometry, though not specifically taught for that purpose, **supplies an integration** of much of mathematics. Pictures, graphs, areas under a curve, slopes of lines, sets of points, and many other devices are used to help students visualize mathematical concepts. "A picture is worth a thousand words."

A few students (too few) have said the reason they were taking geometry was because they thought it would be **interesting and enjoyable.** These are usually students who enjoy thinking and enjoy games and puzzles. Indeed, much of geometry, such as topology and combinatorial geometry, has its origin in problems and puzzles. Some of the more famous are:

*The Seven Bridges of Königsberg*—In the eighteenth century, the city of Königsberg (now Kalingrad in the U.S.S.R.) had seven bridges situated in the following manner:

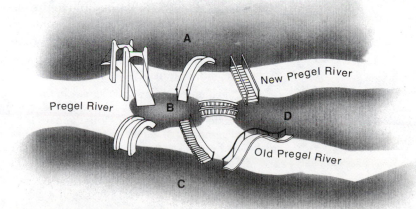

The bridges connected the two banks of a river and two islands in the river. The problem is: can a walker cross all seven bridges without crossing any one of them twice? He may start anywhere and end anywhere.

A geometrically similar and more recent problem concerns three houses, $H_1$, $H_2$, and $H_3$, and three service companies, $S_1$, $S_2$, and $S_3$. The problem is to connect each of the services to each of the houses.

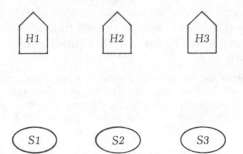

No pipe (line) is allowed to cross another or to pass under a house.[1]

An older and more serious problem was the following: during a plague in Athens in about the year 430 BC (believed now to have been a typhoid epidemic), the people sought the advice of the oracle at Delos as to how they might appease the gods and halt the plague. The oracle replied that if they would construct a cubical altar exactly twice the size of the present cubical altar, the plague would cease. The people made the error of constructing a new altar with an edge twice the original edge (which made the volume eight times as great) and the plague continued to run its course. If the original edge had a length of 1 unit, how long should the new edge have been?

There are many other famous problems, such as the trisection of a general angle with straightedge and compass, and the construction of a square with an area exactly equal to that of a given circle.[2] All of the problems discussed so far are ones that have been *proven* impossible. Euler proved the Königsberg Bridge problem impossible in 1736, and in 1837, Wantzel solved the doubling of a cube problem. (The new edge $\sqrt[3]{2}$ units—which Wantzel showed is not constructible with straightedge and compass.)

What does it mean to say you have "solved" a problem? You must either present a solution (that is, do what the problem requires), or *prove* that what is required is impossible to do. Now, we need to discuss what is meant by "prove". But first, two quotations illustrating the high regard mathematicians have for the subject geometry.

"If Euclid failed to kindle your youthful enthusiasm, then you were not born to be a scientific thinker."—*Albert Einstein*

". . . geometry plays a special part in the curriculum. It seems to be the only mathematical subject that young students can understand and work with in approximately the same way as a mathematician. . . . I hope that we will not allow classical geometry to get lost, unless and until we find material that can take its place. As far as I know, no such material has yet been found."—*Edwin Moise*

---

[1]For a discussion of these problems, and their solution using graph theory, see *Puzzles & Graphs*, by John Fujii, National Council of Teachers of Mathematics; 1966.

[2]For a good article on the squaring of the circle see "Squaring the circle—for fun and profit," *The Mathematics Teacher*; April, 1978.

1–8. Answer as many as possible of the questions asked on pages 3 and 4. You are really not responsible for being able to do these problems at this point, but by the end of the course you should be able to do them all. As a help, here are some of the necessary relationships:

The area of a rectangle is its length times its width.

$$A = lw$$

The volume of a cylinder is $\pi$ times the square of the radius of the base times its height. (Use 3.1416 to approximate $\pi$.)

$$V = \pi r^2 h$$

The area of a trapezoid is half the height times the sum of the bases.

$$A = \frac{h}{2}(b_1 + b_2)$$

The total surface of a cylinder is the sum of the two bases ($2\pi$ times the radius squared) and the lateral surface ($2\pi$ times the radius of the base times the height).

$$S = 2\pi r^2 + 2\pi rh$$
$$\text{or } S = 2\pi r(r + h) \quad \text{(by factoring)}$$

And the Pythagorean triangle relationship: The sum of the squares of the legs of a right triangle equals the square of the hypotenuse.

$$a^2 + b^2 = c^2 \text{ where } c \text{ is the hypotenuse,}$$
that is, the side opposite the right angle.

All of the above relationships will be *proven* later in the course.

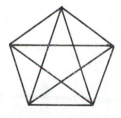

9. Recognition of patterns is important in geometry. This symbol, the star of which was the recognition symbol of the Pythagoreans, contains many triangles. How many are there?

10. The ability to visualize three-dimensional objects (called spatial perception) is also useful in geometry. On a normal die, the sum of the number of spots on opposite faces is always seven; the three dice below are normal in that respect. However, two of the dice below are alike and one is different in respect to the *orientation* of the faces. Which is different?

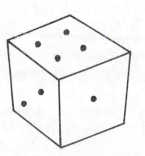
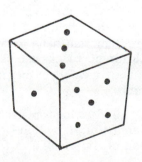
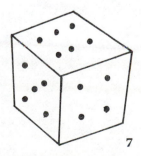

11. In geometry it is useful to be inventive and to look for different approaches to a problem. Connect these nine points using only *four* straight lines. Draw the pattern without lifting the pencil or retracing any of the lines. Here are some attempts using five lines.

(For a method of connecting the dots with *one* straight line see p. 18 of *Conceptual Blockbusting*, by James C. Adams; W. H. Freeman, 1974.)

12. In geometry it is often necessary to be aware of unstated relationships or restrictions. On a regular 8 × 8 checkerboard, 32 dominoes, each the size of two squares, could be used to completely cover the board. If the two opposite corner squares of the board are cut off, arrange 31 dominoes so as to cover the board.

The following problems are called packing problems. They are a part of what is called Combinatorial Geometry.

13. Given a box measuring 3 × 4 × 6 can you pack into it 9 bricks measuring 1 × 2 × 4? You are not allowed to break the bricks! Notice that the volume of the bricks is 72 cubic units and the volume of the box is 72 cubic units, but that does not necessarily mean the bricks will fit into the box.

14. Will two bricks 1 × 6 × 6 fit into the above box?

15. Will four bricks 1 × 3 × 6 fit into the above box?

16. Will eight bricks 1 × 3 × 3 fit into the above box?

17. Will 27 bricks 1 × 2 × 4 fit into a box measuring 6 × 6 × 6?

*18. Will 6 bricks 1 × 2 × 2 and 3 bricks 1 × 1 × 1 fit into a box 3 × 3 × 3? (If you analyze this properly, the problem becomes very easy.)

**19. Will 13 bricks 1 × 2 × 4 and 1 brick 2 × 2 × 2 and 3 bricks 1 × 1 × 3 and 1 brick 1 × 2 × 2 fit into a box measuring 5 × 5 × 5? (For this problem, you need to construct a model. It makes a very nice puzzle. Children's blocks can be glued together to make the bricks, they can be cut out of wood, or they can be assembled from snap-together centimeter cubes.)

(A good article on packing problems can be found in *Mathematical Gems, Vol. II*, by Ron Honsberger; published by Mathematical Association of America.)

\*20. Although trial and error is a perfectly legitimate mathematical method, we would like to develop the ability to analyze a problem and make our trial and error more efficient. This problem forces us to analyze and learn from our unsuccessful trials:

> Write a ten digit number in which the first digit gives the number of zeros in the number, the second digit gives the number of ones, the third digit gives the number of digits that are twos, and so on, the tenth digit giving the number of nines contained in the number.

## 2. What is a proof; Deductive reasoning

Ask a hundred mathematicians what constitutes a proof, and you'll probably get a hundred different responses. Indeed, one of the basic problems of mathematics is that we are not in agreement as to what constitutes a proof. Often we dismiss as "obvious," things that to others are not so "obvious." Rather than attempt a careful definition of "proof," we will take the elementary position that **a proof is some demonstration that convinces the observer.** By this definition a magician can *prove* to a small child that his ears are full of quarters, or a television commercial can prove to the viewer that one brand of soap is better than another. Although these may be proofs for some people, they will not be proofs to a more careful observer, or to a better educated person. Consider the following "proof": Let $a$ and $b$ be two names for the same number, that is $a = b$ then

| | |
|---|---|
| $a^2 = ab$ | multiplication of both sides by $a$ |
| $a^2 - b^2 = ab - b^2$ | subtraction from both sides of $b^2$ |
| $(a - b)(a + b) = b(a - b)$ | factoring |
| $a + b = b$ | dividing both sides by $(a - b)$ |
| $b + b = b$ | substitution (remember $a = b$) |
| $2b = b$ | addition |
| $2 = 1$ | dividing both sides by $b$ |

If you remember your algebra, it will seem that every step is legitimate and we have proven that $2 = 1$. Are you convinced? So, we have the double problem that people are sometimes convinced when they should not be, and are sometimes not convinced when they should be. (Although, in this algebraic example, there is the error of division by zero when we divided both sides by $a - b$.)

How can we possibly accomplish anything in science, law, diplomacy, or countless other endeavors, if there is no agreement as to what is a proof? What can we use as reasons to convince others?

What sort of reasoning does a child use? Usually something based on emotion or "authority"; "Because," "It's not fair," "Teacher said," "My daddy said so." Adults also use reasons based on emotion and "authority": "It just isn't right," "He's a better man for the job," "I saw it in the paper," or even "I learned it in my geometry class."

The scientist employs, and finds acceptable for most purposes, a type of reasoning called **inductive logic.** This is a **reasoning from the specific to the general.** It involves recognizing patterns, and then formulating a general rule. Unfortunately, this does not always work. Consider Ptolemy, the second century AD astronomer and mathematician, who very carefully watched the stars and formed the explanation of how they all revolved about the Earth. As better tools for observation became available, deviations in the paths of planets were observed, and

Ptolemy's theories needed more and more revision and modification. Finally, Copernicus, early in the sixteenth century, presented a whole new theory based on the belief that the Earth and other planets revolved about the sun. Another, more recent, example of how conclusions based on observations are discarded when more precise observations are made can be found in modern nuclear physics. It wasn't too many years ago that the proton, neutron, and electron were the basic building blocks of matter. Now there are so many subatomic particles being discovered that general theories that explain all the phenomena are nearly impossible to formulate. Inductive reasoning is based on observed patterns, but the next item in a sequence may not be what you expected—you may be using a different rule than the generator of the pattern. Consider the sequence 2, 4, _____. What is the next number? Obviously, it depends upon the rule. Several persons can arrive at several numbers by using different rules. Try a longer sequence: 1, 8, 15, 22, 29, _____. The next number is not necessarily 36. You used inductive reasoning in observing, generalizing, and then applying your generalization. If you generalized that each number was 7 greater than the previous you were incorrect. The next number should be 5, as these are the dates of Tuesdays beginning January 1, 1980. And what does one do when there doesn't seem to be a pattern? Much of the world around us seems to be patternless. Does this mean there is no pattern or plan? Or is it perhaps just too complex for us to figure out? Sometimes we have pre-set notions that prevent us from seeing even simple patterns. This pattern appeared on a national mathematics exam:

| 7 | 8 | 9 | 10 |
|----|----|----|----|
| 94 | 46 | 18 | ? |

(Children seem to find it easier than adults!)

What is the next element in each of these sequences?

o, t, t, f, f, s, s, e, _____

2, 4, 16, 37, 58, 89, 145, 42, 20, _____

These are difficult so don't spend *too* much time on them, but consider this quote from the famous mathematician and problem-solver, George Polya:

A great discovery solves a great problem, but there is a grain of discovery in the solution of any problem. Your problem may be modest; but if it challenges your curiosity and brings into play your inventive faculties, and if you solve it by your own means, you may experience the tension and enjoy the triumph of discovery. Such experiences at a susceptible age may create a taste for mental work and leave their imprint on the mind and character for a lifetime.

**DEDUCTIVE REASONING**

There is one form of reasoning that is infallible. (Well, almost.) The reasoning that a mathematician or a logician uses for proofs is called **deductive reasoning, which proceeds from a general, accepted statement to a specific case.**

With the use of deductive reasoning we can construct **proofs** of statements and heed Buddha's parting injunction to his followers: "Believe nothing on hearsay. Do not believe in traditions because they are old, or in anything on the mere authority of myself or any other teacher."

Deductive reasoning is based on the use of conditional statements. **Conditional statements are ones that are, or can be, put into an if–then form.** If it is a square, then all of its sides are equal. Water is wet (if it is water, then it is wet). If it is snowing, then I'll get a "B" in Geometry. If the "if" part of a conditional statement is true, then the "then" part must also be true. Venn diagrams (sometimes called Euler circles) are useful to show this relationship.

Venn diagrams use circles to represent sets. Everything placed inside the circle belongs to the set, that is, it has the properties of the set, and everything outside of the circle does not belong to the set.

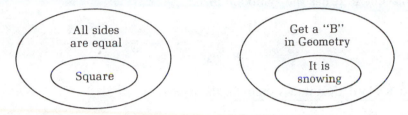

The "if" part of a conditional statement is called the **antecedent** or the **hypothesis,** and the "then" part is called the **consequent** or the **conclusion.** If the hypothesis is represented by the variable $A$, and the conclusion by the variable $B$, then the statement "if $A$, then $B$" can be represented symbolically by $A \rightarrow B$, and the Venn diagram is:

The "if" is a subset of the "then". If a conditional statement is accepted, and the hypothesis is true, then the conclusion is inescapable—it must be true. The **reasoning is valid** if this form is followed. In formal logic this is called the **method of affirming the consequent by affirming the antecedent.** To state symbolically that "if $A$ is true then $B$ is true; $A$ is true, therefore $B$ must be true," we write:

$$A \rightarrow B$$
$$\underline{A}$$
$$\therefore B$$

If I have a flat tire, then I'll be late for class.
    *A*                   *B*

I did have a flat tire.
    *A*

Therefore, I will be late for class.
       *B*

$A \rightarrow B$

$A$

$\therefore B$

If the toothpicks are missing, then your pet goat was here.
    *M*                  *G*

The toothpicks are missing.
    *M*

Therefore, your pet goat was here.
       *G*

$M \rightarrow G$

$M$

$\therefore G$

| If Aussie Airlines flies to Australia, then the koala bears will leave. | $A \rightarrow B$ |
|---|---|
| Aussie Airlines does fly to Australia. | $A$ |
| Therefore, the koala bears will leave. | $\therefore B$ |

Notice that we are interested only in *valid reasoning*, and what the statements say is unimportant!

| If I am a qwert, then I will yuiop. | $Q \rightarrow Y$ |
|---|---|
| I am a qwert. | $Q$ |
| Therefore, I will yuiop. | $\therefore Y$ |

You should probably be warned at this point of a common error in the use of deductive reasoning. It is called the **fallacy of affirming the consequent.** It has the symbolic form:

$$A \rightarrow B$$
$$B$$
$$\therefore A$$

| If it is a bird, then it flies south for the winter. | $B \rightarrow S$ |
|---|---|
| ⎵$B$　　　　　⎵$S$ | |
| It does fly south for the winter. | $S$ |
| ⎵$S$ | |
| Therefore, it is a bird. | $\therefore B$ |
| ⎵$B$ | |

This can be seen as a fallacy by looking at the Venn diagram.

Things that fly south for the winter

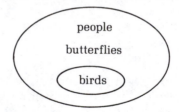

people

butterflies

birds

Notice that there are many other things that go south for the winter, and that this property will not guarantee it is a bird.

There is another method of using a conditional statement in order to draw a valid conclusion. This is called the **method of denying the antecedent by denying the consequent.** It has the symbolic form:

$$P \rightarrow Q$$
$$\sim Q$$
$$\therefore \sim P$$

where $\sim$ is read as "not". In words: "If $P$, then $Q$, and not $Q$, therefore not $P$." We can see from the Venn diagram that not $Q$ must lie outside of the larger set and could certainly not be $P$.

Some examples:

If that is a rock, then it must be hard.      $R \to H$
      R             H
It is not hard.      $\sim H$
  H
Therefore, it is not a rock.      $\therefore \sim R$
    R

If he is Mr. Spock, then he is an alien.      $S \to A$
    S         A
He is not an alien.      $\sim A$
  A
Therefore, he is not Mr. Spock.      $\therefore \sim S$
    S

Once again remember that this is a method of **valid deductive reasoning**, and we are interested only in the form of the statements and not what they say.

There is a common error in the use of this form of reasoning also. It is called the **fallacy of denying the antecedent.** It has the symbolic form:

$$P \to Q$$
$$\underline{\sim P}$$
$$\therefore \sim Q$$

If you get an "A", then you must be smart.      $A \to S$
   A          S
You did not get an "A".      $\sim A$
   $\sim A$
Therefore, you must not be smart.      $\therefore \sim S$
   $\sim S$

Again the Venn diagram is useful in showing that this is a *fallacy*, and not valid reasoning.

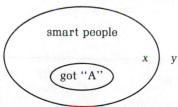

If you did not get an "A", you could be either x or y, which says you could be either smart or not smart. The conclusion that you are not smart is not justified. You should be familiar with the two forms of valid deductive reasoning and be aware of the two common fallacies.

One trap that people sometimes fall into is assuming the hypothesis has been met when in fact it has not.

**All senior girls will get a letter from Planned Parenting.**
**Jean is a senior.**
**Therefore, Jean will receive a letter from Planned Parenting.**

As it turns out, in this case Jean is a boy; so the hypothesis has not been met, and the conclusion is not justified.

**All senior citizens ride the bus free.**
**Grandma is 66.**
**Therefore, Grandma can ride the bus free.**

Perhaps not—what is the bus line's definition of "senior citizen"? Has the hypothesis really been met? For Grandma's sake, we hope so!

Since not all reasoning can be put into neat little conditional statements, it might be good to be aware of some common abuses of everyday reasoning.

**Change of meaning:** "I am a small person, my wife is a small person, and all of our seven children are small. Therefore, we are a small family."

**Diversion** (stating a fact that really does not prove the point): "Which of these three airlines has the cheapest fare—Aussie Airlines, TWP, or TransAff? As a matter of fact, TransAff has a 35 per cent discount on its night coach."

**Reliance on "experts,"** who may not be experts in the proper field. This includes such tactics as using an all-star football player to sell lingerie (Does he wear it to play football?), and getting a popular singer to sell saltine crackers (but not while he has a mouth full of them).

**Pleasure association** (claiming certain pleasurable situations are connected to a particular product or activity, usually status, attractiveness to the opposite sex, or financial success): "Triple your pleasure, triple your fun, with Triplespear Gum." "He drives a DeLuxe, and she loves him." "Fly TransAff—we move our tails for you." (This last is also an example of change of meaning.) "If you are an Irish Brand Brandy man, you'll have the sweet smell of success."

**Bandwagon effect** (everyone does it, it must be right): "More people prefer Tester Toasted Coffee." "Buyer's Children's Aspirin—the pain reliever more mothers prefer."

**Loaded words** (words with emotional overtones): This includes a hamburger stand that advertises a "super-taco." (Does it fly?)

Watch for political use of emotionally loaded words: feminist, foolish, spendthrift, conservative, liberal, and other words that are only the speaker's opinion. Learn to analyze the logic in statements and arguments, and not just react emotionally to the speaker.

**EXERCISE SET 2**

1–25. Classify each of the following arguments as valid or not valid reasoning. The "truth" of the statements is unimportant; consider only the pattern of reasoning.

1. Enlisted men in the Navy must be at least 1.6 m tall.
   George is an enlisted man in the Navy.
   Therefore, George is at least 1.6 m tall.

2. Good pianists give recitals.
   Karen is giving a recital.
   Therefore, Karen is a good pianist.

3. To be eligible for baseball, you must pass ten units.
   Frank passed ten units.
   Therefore, Frank is eligible for baseball.
   (It might be helpful to rewrite the conditional statement in if-then form.)

4. Freshman girls are innocent and pure.
   Mary is not a freshman girl.
   Mary is not innocent and pure.

5. Senior boys are not innocent and pure.
   Paul is innocent and pure.
   Paul is not a senior boy.

6. All churches are painted white.
   This building is painted white.
   This building is a church.

7. Children are admitted free to the movies.
   Karl is 14 years old.
   Karl will be admitted free to the movies.

8. All churches are painted white.
   This building is a church.
   This building is painted white.

9. Adults are not admitted free to the movies.
   Keith was admitted free.
   Keith is not an adult.

10. If it is Colombian, then it is great.
    This is great.
    It is Colombian.

11. Ten thousand children were inoculated with anti-paralysis
    vaccine, and none of them had infantile paralysis.
    Thomas did not have infantile paralysis.
    Thomas was one of the children inoculated.

12. If a closed plane figure has four sides, then it is a quadrilateral.
    A quert has four sides.
    A quert is a quadrilateral.

13. If a closed plane figure has four sides, then it is a quadrilateral.
    A rectangle is a quadrilateral.
    A rectangle has four sides.

14. All great men have poor penmanship.
    I am not a great man.
    I do not have poor penmanship.

15. If a triangle has three equal sides, then it is equilateral.
    This triangle has three equal sides.
    This triangle is equilateral.

16. All rectangles are parallelograms.
    This figure is a rectangle.
    This figure is a parallelogram.

17. Children between 8 and 16 years of age must attend school.
    Jonathan attends school.
    Jonathan is between 8 and 16 years of age.

18. All Texans are tall.
    Gary Cooper is tall.
    Gary Cooper is a Texan.

19. Dictionaries contain definitions of words.
    This book contains definitions of words.
    This book is a dictionary.

20. Dictionaries contain definitions of words.
    This book does not contain definitions of words.
    This book is not a dictionary.

21. Dictionaries contain definitions of words.
    This book is not a dictionary.
    This book does not contain definitions of words.

22. If it is Colombian, then it is great.
    This is not Colombian.
    This is not great.

23. No students over 80 ride bicycles to school.
    Irving is not over 80.
    Irving does not ride a bicycle to school.

24. If it is green, then it is a frog.
    This is not green.
    This is not a frog.

25. If I am tired, then I cannot do my homework.
    I can do my homework.
    Therefore, I am not tired.

26–41. Classify each of the following abuses of logic as change of meaning, diversion, reliance on experts, pleasure association, bandwagon, or loaded words (more than one classification may apply).

26. More Californians choose Bank of Calamer than any other bank.

27. If each of these three airlines—Aussie Airlines, TWP, and TransAff—leaves for Chicago at the same time, which one will get there first? TWP has the best on-time record as shown in the records of the Civilian Aeronautics Bureau.

28. Marie Antoinette recommends Pierre's French Cakes.

29. He's a real card. He needs to be dealt with.

30. Get your next rental car from us. We try harder.

31. He's one of the big men at city hall.

32. Lucille Ball is Sonny Bono's mother; I read it in the weekly *National News*.

33. I thoroughly oppose my opponent's left-wing radical principles.

34. Of course I pop my gum in class, everybody does it.

35. Would you really vote for that pro-abortionist baby-killer?

36. Croak gives life—every body wants a little life.

37. J. C. was a good governer; he will be a good president.

38. Only bleeding heart liberals are against capital punishment.

39. Buyer's Children's Aspirin—the pain reliever more mothers prefer.

40. We don't sell tacos—we sell a "Super-Taco."

41. I'm country singer Nat Boon, and I always eat Ambisco Brand Saltines.

42–44. Find three examples of abuses of logic. (Television commercials and political speeches are excellent sources.) Identify each example as change of meaning, diversion, reliance on experts, pleasure association, bandwagon, or loaded words, or any combination of these.

45–48. Make up your own examples illustrating each of the two valid methods of deductive reasoning and each of the two fallacies.

## 3. Syllogisms

The preceding examples of deductive reasoning probably seem pretty trivial and hardly suitable for the foundation of scientific reasoning. What we need to do now is put together strings of several conditional statements in such a manner that the results can be a little more rewarding. The process is very similar to the transitive property that you used in algebra (if $a = b$ and $b = c$, then $a = c$). As an example: The whale is a mammal. All mammals are warm-blooded. Therefore, the whale is warm-blooded. Rephrasing these statements in the if-then form: If it is a whale, then it is a mammal. If it is a mammal, then it is warm-blooded. Therefore, if it is a whale, then it is warm-blooded. The symbolic logic would be:

$$
\begin{array}{l} A \to B \\ B \to C \\ \hline \therefore A \to C \end{array}
\qquad \text{or} \qquad
\begin{array}{l} B \to C \\ A \to B \\ \hline \therefore A \to C \end{array}
$$

(The order in which the first two statements are given is unimportant.)

In terms of the Venn diagram: If $A$ is a subset of $B$, and $B$ is a subset of $C$, then $A$ is a subset of $C$:

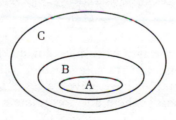

Our example becomes:

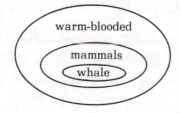

This technique of creating a new conditional statement from two others is called a *syllogism*. The first statement is called the major premise, the second statement is called the minor premise, and the third statement is the conclusion. A few more examples:

| | |
|---|---:|
| If I want to get that job, then I'll have to get my degree. | $J \to D$ |
| If I am to get my degree, then I'll have to pass geometry. | $D \to P$ |
| Therefore, if I want to get that job, I'll have to pass geometry. | $\therefore J \to P$ |

| | |
|---|---:|
| If the lines are perpendicular, they form right angles. | $P \to R$ |
| All right angles are equal. | $R \to E$ |
| Therefore, if the lines are perpendicular, the angles are equal. | $\therefore P \to E$ |

| | |
|---|---:|
| If he is the best man, then most people will vote for him. | $B \to V$ |
| If most people vote for him, then he will win. | $V \to W$ |
| Therefore, if he is the best man, then he will win. | $\therefore B \to W$ |

| | |
|---|---:|
| If I miss my dinner, my stomach will growl. | $M \to G$ |
| If I work too late, I will miss my dinner. | $W \to M$ |
| Therefore, if I work too late, my stomach will growl. | $\therefore W \to G$ |

This transitive type property, just like an algebraic property, can be extended to the use of more than just two conditional statements. We can put together strings of any number of conditional statements to arrive at a valid conclusion in the form of a new conditional statement. As an example:

| | |
|---|---:|
| If I continue at my part-time job, I can finance college. | $C \to F$ |
| If I can finance college, I can get a degree. | $F \to D$ |
| If I can get a degree, I will earn more money. | $D \to \$$ |
| If I earn more money, I will get married. | $\$ \to M$ |
| If I get married, I will become a parent. | $M \to P$ |
| Therefore, if I continue at my part-time job, I will become a parent. | $\therefore C \to P$ |

The Venn diagram, showing the nested subsets, makes the conclusion obvious:

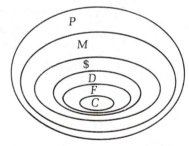

Unfortunately, the information we need for making decisions (conditional statements) is not always so available, well stated, and well organized as in the example above. Given a set of statements, it makes a nice little game to reorder them and formulate a new valid statement.

| | |
|---|---:|
| If I get my hands dirty, I will wipe them on my pants. | $D \to W$ |
| If I drive 3000 km, I will check my oil. | $3 \to C$ |
| If I wipe my hands on my pants, it will ruin them. | $W \to R$ |
| If I check my oil, I will get my hands dirty. | $C \to D$ |

Symbolically, the preceding four statements should be rearranged as:

$$3 \to C$$
$$C \to D$$
$$D \to W$$
$$\underline{W \to R}$$
$$\therefore 3 \to R$$

And we can conclude: If I drive 3000 km, I will ruin my pants.

Reordering well stated statements is not too difficult, since the antecedent of our new statement is the only antecedent that doesn't occur as a consequent, and the consequent of the new statement is the only consequent that doesn't occur as an antecedent in our original statements. (In our example, the 3 does not occur in the right-hand column, and the R does not occur in the left-hand column.)

Often (in fact, usually), statements are not nice, neat, if-then sentences. We therefore need to learn to recognize and manipulate other forms; in particular, three variations of the conditional statement called the converse, the inverse, and the contrapositive.

If we start with the statement $A \rightarrow B$, then the **converse is formed by interchanging the hypothesis and the conclusion:** $B \rightarrow A$.

> Statement:  $A \rightarrow B$
> Converse:   $B \rightarrow A$

These two statements do not say the same thing, although in some instances they might both be accepted as true. Examples:

**If the triangles are congruent, then $\sphericalangle 1 = \sphericalangle 2$.**          $C \rightarrow =$
**If $\sphericalangle 1 = \sphericalangle 2$, then the triangles are congruent.**          $= \rightarrow C$

**If it is Tuesday, I go to geometry.**          $T \rightarrow G$
**If I go to geometry, it is Tuesday.**          $G \rightarrow T$

**If he is drunk, he can't touch his nose.**          $D \rightarrow T$
**If he can't touch his nose, he is drunk.**          $T \rightarrow D$

**If I am elected, your taxes will be lower.**          $E \rightarrow L$
**If your taxes are lower, then I was elected.**          $L \rightarrow E$

Often when a statement is made, people incorrectly assume the converse. Example: If you do well on your midterm, you will get an "A". Assume now that you got an "A". Does that mean that you did well on your midterm? Perhaps not. Look at the Venn diagram:

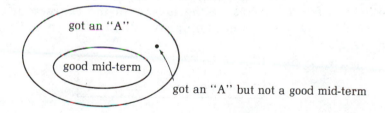

got an "A" but not a good mid-term

Suppose you tell your child: "If you clean your plate, you will get dessert." The child will probably assume that if he is to get dessert, all of his vegetables will have to be eaten. But that is not what was said. Consider the Venn diagram. There are people who get dessert who did not clean their plates. (Besides, Mom is softhearted.)

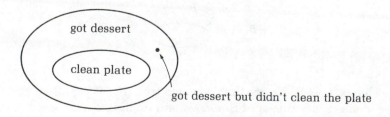

got dessert but didn't clean the plate

It should be mentioned that when a statement uses the expression "if and only if" (abbreviated iff), it then means that the statement and its converse must both be true. This is the case with all definitions.

If we start with the statement $A \rightarrow B$, then the **inverse is formed by negating both the hypothesis and the conclusion:** $\sim A \rightarrow \sim B$.

> Statement:  $A \rightarrow B$
> Inverse:  $\sim A \rightarrow \sim B$

The symbol $\sim$ can be read as "not." (Another common notation for $\sim A$ is $A'$.) Again, these two statements do not have the same meaning. Some examples:

| | |
|---|---|
| **If it is snowing, then it is cold.** | $S \rightarrow C$ |
| **If it is not snowing, then it is not cold.** | $\sim S \rightarrow \sim C$ |
| **If it is a rectangle, then it is a parallelogram.** | $R \rightarrow P$ |
| **If it is not a rectangle, then it is not a parallelogram.** | $\sim R \rightarrow \sim P$ |

Assuming the inverse of a statement is also a common mistake: "If I am elected, we will have lower taxes." Some people will assume that if the speaker is not elected, we will not have lower taxes. "If you are absent six times, you will flunk the course." Would you assume that if you are not absent six times, you will not flunk the course?

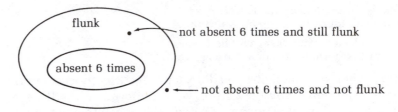

The most useful of the alternate forms of a conditional statement is the contrapositive. If we start with the statement $A \rightarrow B$, then the **contrapositive is formed by taking the inverse of the converse** (or the converse of the inverse). Sounds impressive, doesn't it? Symbolically it is $\sim B \rightarrow \sim A$.

> Statement:  $A \rightarrow B$
> Contrapositive:  $\sim B \rightarrow \sim A$

Some examples:

| | |
|---|---|
| **If it is isosceles, it has two equal legs.** | $I \rightarrow L$ |
| **If it does not have two equal legs, it is not isosceles.** | $\sim L \rightarrow \sim I$ |
| **If I read the book, then I can do the homework.** | $R \rightarrow H$ |
| **If I cannot do the homework, then I did not read the book.** | $\sim H \rightarrow \sim R$ |

Let's look at the Venn diagram:

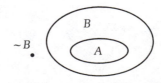

This shows $A \rightarrow B$. Notice that it is also true that if $\sim B$, then certainly $\sim A$—since $A$ is a subset of $B$ and we are outside of $B$.

Any statement and its contrapositive say exactly the same thing, but in two different ways. You can replace one with the other any time it is convenient to do so.

In order to further manipulate statements, we need to be aware of the fact that the negation of a negation is the original. That is $\sim(\sim A)$ is the same as $A$. If it is untrue that the lawn is not green, then the lawn is green. If something is said not to be unlikely, then it is likely. Let us use this concept in the writing of some contrapositives:

| | |
|---|---|
| **If I watch that program, I will not get to bed on time.** | $W \to \sim B$ |
| **If I get to bed on time, I will not watch that program.** | $\sim(\sim B) \to \sim W$ |
| | or $B \to \sim W$ |
| | |
| **If you do not water the flowers, they will die.** | $\sim W \to D$ |
| **If your flowers did not die, then you watered them.** | $\sim D \to \sim(\sim W)$ |
| | or $\sim D \to W$ |

Statements using the words "none" and "no" are sometimes difficult to rephrase into an if–then form. The Venn diagram will help. What a "none" or "no" says is that two sets are **disjoint sets.** That is, any member of one set cannot be a member of the other. Disjoint sets are represented by two separate circles. No $A$'s are $B$'s:

| | |
|---|---|
| This says: If $A$, then not $B$. | $A \to \sim B$ |
| Or, by contrapositive: If $B$, then not $A$. | $B \to \sim A$ |

**Example: No parallel lines intersect.**

| | |
|---|---|
| **If lines are parallel, then they do not intersect.** | $P \to \sim I$ |
| **Or: If lines intersect, then they are not parallel.** | $I \to \sim P$ |

**Example:**
**None of the professors come to work on a skateboard.**

| | |
|---|---|
| **If he or she is a professor, then he or she does not come to work on a skateboard.** | $P \to \sim S$ |
| **If he or she comes to work on a skateboard, then he or she is not a professor.** | $S \to \sim P$ |

Now let's use double negation, contrapositives, and a string of statements to form a new statement:

**If I'm transferred to Peoria, I'll buy a house.**
**If I'm not transferred to Peoria, then I'll not get a promotion.**
**If I buy a house, I'll have to mow the lawn.**
**If I marry the boss's daughter, I'll get a promotion.**

We will represent these statements by:

$$T \rightarrow H$$
$$\sim T \rightarrow \sim P$$
$$H \rightarrow M$$
$$D \rightarrow P$$

If we take the contrapositive of the second statement and rearrange them, we will have:

$$D \rightarrow P$$
$$P \rightarrow T$$
$$T \rightarrow H$$
$$H \rightarrow M$$

and we can conclude: If I marry the boss's daughter, I'll have to mow the lawn.

**EXERCISE SET 3**

1–6. For each of the following statements write (a) the converse, (b) the inverse, and (c) the contrapositive.

1. If two lines are parallel, the alternate interior angles are equal.

2. If George was guilty, then he was at the scene of the crime.

3. The car will not break down if it has had proper care.

4. If Kathy does not pass the language requirement, she will not graduate.

5. If you have an urge to roll in the mud, you have swine flu.

6. If the sun is shining, then it is not night time.

7–12. Name the relationship (converse, inverse, contrapositive, none) of the second statement to the first.

7. A dog that says "meow" needs a psychiatrist.
   A dog that does not say "meow" does not need a psychiatrist.

8. No triangles are squares.
   No squares are triangles.

9. Class attendance will be down if the surf is up.
   If class attendance is down, then the surf is up.

10. Your teeth will be shiny and white if you eat peanuts.
    If you do not eat peanuts, then your teeth will be shiny and white.

11. People with swine flu run temperatures over 38°C.
    If your temperature is not over 38°C, then you do not have swine flu.

12. Bald headed men are more virile.
    A man is not more virile if he is not bald headed.

13–23. Use each of the following sets of statements to form a new conditional statement. Rearrange, and use contrapositives and double negation where necessary to form a string of conditional statements, where the antecedent of the first statement implys the consequent of the last statement.

13. If it is not a parrallelogram, then it is not a rectangle.
    All squares are rectangles.
    If it is a parallelogram, then it is a quadrilateral.

14. Nothing is better than a new car.
    Riding a bike is better than an old car.
    An old car is better than nothing.
    (Ignore the change of meaning of "nothing.")

15. If you support the church rummage sales, you do not store
    your old clothes.
    All elephants have trunks.
    If you are not small, then you support the rummage sales.
    If you have a trunk, then you store old clothes.
    (Ignore the change of meaning of "trunk.")

16. If the drought continues, water will be rationed.
    If the water department has less revenue, it will have to raise
    the rates to cover expenses.
    If there is water rationing, people will use less water.
    If people use less water, the water department will have less
    revenue.

17. If I break my shoelace in the morning, it will take me ten
    minutes to find another one.
    If I don't get a raise, I won't get married.
    If I take ten minutes to find a shoelace, I will be late for work.
    I will get a raise, if I am not late for work.

18. If we develop a neutron warhead, it will be deployed to the
    NATO forces.
    If other countries do not use atomic weapons, then NATO will
    not use atomic weapons.
    If the neutron warhead is deployed to NATO forces, they will
    use it.
    If Carter approves the research budget, our scientists will de-
    velop a neutron warhead.
    If a nuclear war is started, it will be the end of the world.
    If other countries use atomic weapons, it will start a nuclear
    war.

19. If our youth are well educated, they will become competent
    citizens.
    If we support our schools financially, they will do a good job of
    educating our youth.
    If our technology improves, we will have a higher standard of
    living.
    If our society has competent citizens, our technology will
    improve.

20. If I attend college, I will have to buy a car for transportation.
    If I buy a car, I will have to earn more money.
    If I have to earn more money, I will need a full-time job.
    If I have a full-time job, I will not be able to attend college.

21. If I wash my car, it is sure to rain.
    If it rains, the streets will be muddy.
    If the streets are muddy, my car will be dirtier than it is now.

22. For want of a nail, the shoe was lost.
   For want of a shoe, the horse was lost.
   For want of a horse, the rider was lost.
   For want of a rider, the battle was lost.
   For want of a battle, the kingdom was lost.

23. If the water is not warm, then I will not go swimming.
   If a lot of people go swimming, it will be crowded.
   If the water is warm, a lot of people will go swimming.
   If there is not room to swim, I will drown.
   If it is crowded, there will not be any room to swim.

24. Create an example like the above, using at least four statements.

*25. The rather challenging example that follows is credited to the mathematician Charles Dodgson (1832–1898). He is probably better known as Lewis Carroll, the pen name he used when writing stories for children. Represent each phrase by a variable letter, write each statement symbolically, use contrapositives where necessary, and form a string of conditional statements that lead to a new statement.

**All the dated letters in this room are written on blue paper.**
**None of them are in black ink, except those written in the third person.**
**I have not filed any of those that I can read.**
**None of those that are written on one sheet are undated.**
**All of those that are not crossed out are in black ink.**
**All of those that are written by Brown begin with "Dear Sir".**
**All of those that are written on blue paper are filed.**
**None of those that are written on more than one sheet are crossed out.**
**None of those that begin with "Dear Sir" are written in the third person.**

---

# 1—Summary

Some of the reasons we study geometry:

  Information
  To learn patterns of thought
  It integrates mathematics
  Interest and enjoyment

Methods of reasoning:

  Inductive—reasoning from specific cases to a generalization
  Deductive—reasoning from a generally accepted statement to a specific case

  has the form:  $A \rightarrow B$    (if $A$ then $B$)
  $$\frac{A}{\therefore B}$$    (A is true)
      (therefore $B$ is true)

Common abuses of reasoning:

  Change of meaning
  Diversion
  Reliance on "experts"
  Pleasure association
  Bandwagon effect
  Loaded words

Syllogisms:

Two conditional statements giving a third (conclusion) statement

$$A \rightarrow B$$
$$\frac{B \rightarrow C}{\therefore A \rightarrow C}$$

Converse:

The converse of $A \rightarrow B$ is $B \rightarrow A$
If a statement is true, its converse is not necessarily true.

Inverse:

The inverse of $A \rightarrow B$ is $\sim A \rightarrow \sim B$
If a statement is true, its inverse is not necessarily true.

Contrapositive:

The contrapositive of $A \rightarrow B$ is $\sim B \rightarrow \sim A$
The contrapositive of a true statement is always true.

Strings of conditional statements to create a new statement:

$$A \rightarrow B$$
$$B \rightarrow C$$
$$C \rightarrow D$$
$$\frac{D \rightarrow E}{\therefore A \rightarrow E}$$

The "Thomas Mills house" in Philadelphia, built in 1806, stood on its site for nearly 170 years before being moved one block to make room for hospital construction. Thomas Mills was the first American-born and -trained architect, and this house is the last of his design remaining. The steel I-beams in the foundation were required to support the structure during its journey and will serve as the building's new foundation, without which it would collapse.

# CHAPTER 2

# FOUNDATIONS

In this chapter, we will look at the foundations of the deductive science of geometry: the axioms, the postulates, and the definitions. We will then look at some basic constructions, and use them to consider the different types of transformations. (These transformations will be used later as the basis of the important concepts of congruence and similarity.) We will then look at the format used in the classical form of the formal proof.

*Corollary to Murphy's Law: People disagreeing with your "facts" are always emotional and employ faulty reasoning.*

Now that we have seen how deductive logic can be used to develop new statements from a given set of statements, we can discuss the nature of a deductive science. Notice that in many of the examples used, you could not agree with the original conditional statements and, therefore, did not necessarily agree with the conclusions. In order to develop a deductive science, we need to find a set of original conditional statements with which everyone can agree. Then, if we use the infallible methods of deductive reasoning, we can develop new statements (new knowledge) that will be valid. These **statements that are assumed, or accepted without proof,** are called **axioms and postulates.**

Without going into the philosophical considerations of what is "truth," let us claim that all knowledge and "truth" is founded on some basic assumptions that were *not* proved. In mathematics, or any other deductive science, these assumptions are stated and are called axioms or postulates. The not-too-important distinction between axioms or postulates is that axioms pertain to all mathematics, and postulates pertain to the specific topic being studied. In this course, our postulates will be

**4. Axioms**

statements that deal with geometrical concepts. The axioms will be statements that are also applicable to other areas of mathematics.

Even more basic than the axioms and postulates is the necessity for agreement on **definitions** of terms. When we discuss a plane, the student should not be thinking of flying. Or perhaps believing that coincide is what you do when it rains. Or that a polygon is a dead parrot. (Pretty bad?)

The student who wishes to do well in geometry will learn thoroughly the definitions, axioms, and postulates—as they are the only tools we have to begin proving new statements. Statements that we develop and prove are called **theorems.** They are not assumed, but **are proven with the use of definitions, axioms, postulates, and other previously proven theorems.**

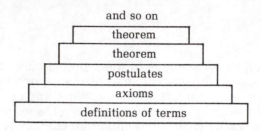

Now for a little story that might help illustrate the need for axioms.[1] Imagine that the annual flooding of the Nile has deposited its rich topsoil for the farmers, but has also erased the boundary lines between neighboring farms, and that Brown and Jones (not very good Egyptian names) are two farmers who, since human nature was probably not very different then, are arguing about the boundary line between their farms. There is a rock at *A* on the river bank and the boundary is supposed to be at right angles to the bank at *A*. Brown claims *AB* is the boundary and Jones holds out for *AC*. How are they to resolve the dispute? A first method would be for one to kill the other and occupy both farms. This method, while fashionable among great nations, does not seem fair when applied to individuals. A second method would be to apply to the King or the High Priest for an arbitrary ruling.

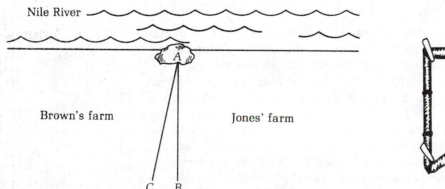

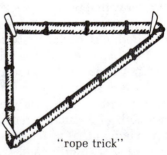

"rope trick"

Both of these methods doubtless enjoyed great prestige in the past and, to a slightly lower degree, still do today, but they do not satisfy a reflective person. Therefore, Brown and Jones try to come to some common agreement as to where the perpendicular goes.

---

[1]*Fundamentals of Mathematics*—M. Richardson

Jones' argument runs something like this: "The boundary is $AC$ because I well remember that in the middle of that lake of mud, exactly at $C$, there used to be a bush beside which my favorite dog used to sleep, and I am ten years older than you, and, even if you did go to school in Alexandria, I've never yet admitted that anyone's judgment was as good as mine, and I'm not going to begin at my time of life, by Isis!"

Brown's argument is that if you take a rope with knots at equal intervals, and stretch it taut about three stakes so that the sides contain 3, 4, and 5 intervals, respectively, then the angle opposite the longest side has to be a right angle (this was probably known to the Egyptian surveyors, who were called "rope-stretchers"); and he had done this and found the boundary to be $AB$. The following dialogue takes place:

Jones:   I don't believe that 3–4–5 hocus-pocus.

Brown:   Do you believe that if the lengths $X$, $Y$, and $Z$ of the three sides of a triangle satisfy the equation $X^2 + Y^2 = Z^2$, then the angle opposite $Z$ is a right angle?

Jones:   If that were true, your 3–4–5 rope trick would be correct, since $3^2 + 4^2 = 5^2$, and you would win the argument; so I don't believe it.

Brown:   Do you believe that if $X$, $Y$, and $Z$ are the lengths of the sides of a right triangle, $Z$ being opposite the right angle, then $X^2 + Y^2 = Z^2$? (the Pythagorean Theorem)

Jones:   That sounds too much like your previous statement. I don't believe it.

Brown:   Do you believe that if two triangles are similar, then their sides are in proportion?

Jones:   If it gives you the land, I don't believe it!

Brown, seeing what he has to contend with, takes a breath and begins again:

Brown:   Do you believe that through two points, one and only one line can be drawn?

The "rope-stretchers" were the surveyors of ancient Egypt. They formed right angles with a rope divided by knots into a ratio of 3:4:5. *The Metropolitan Museum of Art.*

*Jones:* Certainly, but what has that to do with the case?

*Brown:* Do you believe that if equals are added to equals, the results are equal?

*Jones:* Of course. But keep to the subject of this boundary—I'm a busy man.

However, Brown keeps asking Jones whether he believes certain simple statements which happen to be the axioms and postulates of geometry. Jones, impatient and still failing to see any way in which those statements can cause him to lose the argument, admits that he believes them all. Thereupon, Brown proceeds to prove, step by step, on the basis of these axioms and with deductive logic, that the converse of the Pythagorean Theorem is valid and, therefore, that the rope-stretching trick is legitimate, and that the disputed land is his.

So you see that the problem was to agree on some basic principles and then to reason logically from them. Indeed, *nothing* can be proven without resorting to the use of some basic assumptions. As Tennyson puts it in *The Ancient Sage:*

> Thou canst not prove the Nameless, O my son,
> Nor canst thou prove the world thou movest in,
> Thou canst not prove that thou art body alone,
> Nor canst thou prove that thou art spirit alone,
> Nor canst thou prove that thou art both in one:
> Thou canst not prove thou art immortal, no
> Nor yet that thou art mortal—nay my son,
> Thou canst not prove that I, who speak with thee,
> Am not thyself in converse with thyself,
> For nothing worthy proving can be proven,
> Nor yet disproven: wherefore thou be wise,
> Cleave ever to the sunnier side of doubt,
> And cling to Faith beyond the forms of Faith!

So let's get to these statements we take on faith. Most of the axioms you have encountered in algebra—but perhaps have not considered them as having broader application.

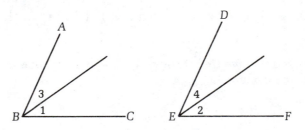

---

    1. *If equals are added to equals the sums are equal.*

---

**Examples:**

$$X = 17$$
$$\therefore X + 3 = 20$$

$$\angle 1 = \angle 2$$
$$\angle 3 = \angle 4$$
$$\therefore \angle ABC = \angle DEF$$

(∢ **is the symbol for an angle, which we will define carefully in the next section, but use your intuitive idea of angle here.**)

2. If equals are subtracted from equals, the differences are equal.

**Examples:**

$$2X - 15 = 27$$
$$5 = 5$$
$$\therefore 2X - 20 = 22$$

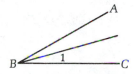

$$AB = CD \qquad AX = CY \qquad so\ XB = YD$$

3. If equals are multiplied by equals, the products are equal.

**Examples:**

$$3A = 5 \qquad \angle 1 = \angle 2$$
$$7 = 7 \qquad \textbf{both are doubled}$$
$$\therefore 21A = 35 \qquad \therefore \angle ABC = \angle DEF$$

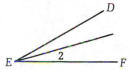

4. If equals are divided by non-zero equals, the quotients are equal.

**Examples:**

$$2X = 10 \qquad \angle ABC = \angle DEF$$
$$2 = 2 \qquad \textbf{both are bisected (divided into two}$$
$$\textbf{equal parts)}$$
$$\therefore X = 5 \qquad \therefore \angle 1 = \angle 2$$

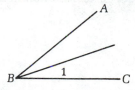 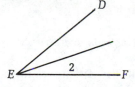

5. Like powers of equals are equal.

**Examples:**

$$T = 9 \qquad A = 2$$
$$\therefore T^2 = 81 \qquad \therefore A^5 = 32$$

6. Like positive roots of equals are equal.

**Examples:**

$$X^3 = 125$$
$$\therefore X = 5$$

$$Y^2 = 169$$
$$\therefore |Y| = 13$$

7. A quantity may be substituted for its equal in any expression.

**Examples:**

$$X + Y = 57$$
$$Y = 13$$
$$\therefore X + 13 = 57$$

$$AB + BC = AC$$
$$AB = CD$$
$$\therefore CD + BC = AC$$

A    B              C    D

8. The whole of a quantity is equal to the sum of its parts.

**Examples:**     $$AB + BC = AC \qquad \sphericalangle DOX + \sphericalangle XOG = \sphericalangle DOG$$

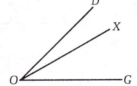

9. All the properties of the real numbers (which we will not review).

We will also find it useful to use some corollaries of the axioms. A **corollary to an axiom, postulate, or theorem is a statement that is easily derived from it.**

1. *Halves of equals are equal.*

A corollary to the division axiom—division by 2.

2. *Doubles of equals are equal.*

A corollary to the multiplication axiom—multiplication by 2.

3. *Quantities equal to the same quantity, or to equal quantities, are equal to each other.*

The substitution axiom used several times.

## 4. *Equal angles have equal complements.*

A corollary to the subtraction axiom. Two angles are **complementary** if the sum of their measures is 90°, or they form a right angle.

$\angle ABC$ **and** $\angle DEF$ **are right angles**
$\angle 1 = \angle 2$
$\angle ABC - \angle 1 = \angle DEF - \angle 2$
$\therefore \angle 3 = \angle 4$

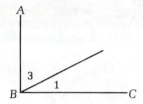 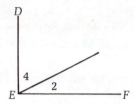

## 5. *Equal angles have equal supplements.*

A corollary to the subtraction axiom. Two angles are **supplementary** if the sum of their measures is 180°, or a straight angle.

$\angle ABC$ **and** $\angle DEF$ **are straight angles**
$\angle 1 = \angle 2$
$\angle ABC - \angle 1 = \angle DEF - \angle 2$
$\therefore \angle 3 = \angle 4$

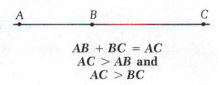

## 6. *The whole of a quantity is greater than any of its parts.*

This is a corollary to Axiom 8, using the subtraction of unequals from equals.

$$\begin{aligned} \text{If}\quad & c = a + b \\ \text{subtracting}\quad & \underline{0 < b} \\ \text{we get}\quad & c > a \end{aligned}$$

**Examples:**

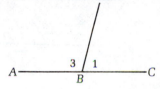 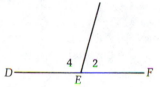

$$AB + BC = AC$$
$$AC > AB \text{ and}$$
$$AC > BC$$

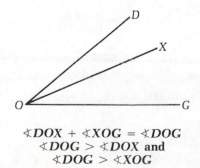

$$\angle DOX + \angle XOG = \angle DOG$$
$$\angle DOG > \angle DOX \text{ and}$$
$$\angle DOG > \angle XOG$$

**EXERCISE SET 4**

1. $7X + 5 = 22$
   $X = \frac{17}{7}$
   State the two axioms applied.

2. $\frac{2}{9}X - 6 = 14$
   $X = 90$
   State the three axioms applied.

3. $\angle X = \angle Y$ and $\angle Y = 45°$
   $\angle X = 45°$
   State one axiom applied.

4. In the adjacent figure we know that
   $DB = EC$
   $AD = AE$
   therefore $DB + AD = EC + AE$
   and $BA = AC$
   State the two axioms applied.

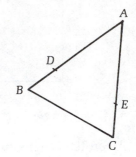

5. In the adjacent figure we know that
   $AB = DC$
   $AE = \frac{1}{2}AB$
   $DF = \frac{1}{2}DC$
   therefore $AE = DF$
   State the corollary applied.

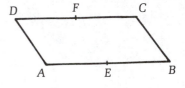

6. In the adjacent figure we know that
   $AD = BC$
   $AF = BG$
   therefore $DF = CG$
   State one axiom applied.

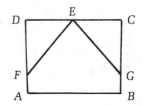

7. In the adjacent figure we know that
   $\angle 1 = \angle 2$
   $\angle 2 = \angle 3$
   $\angle 3 = \angle 4$
   therefore $\angle 1 = \angle 4$
   State one corollary applied.

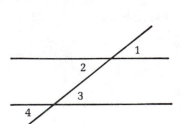

8. In the adjacent figure we know
   $\angle 1 + \angle 2 = 180°$
   $\angle 1 = \angle 3$
   therefore $\angle 3 + \angle 2 = 180°$
   State one axiom applied.

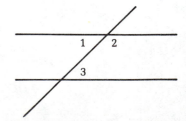

34

9. In the adjacent figure we know that
   $\angle 1 = \angle 2$
   $\angle 1 = \frac{1}{2}\angle A$
   $\angle 2 = \frac{1}{2}\angle B$
   therefore $\angle A = \angle B$
   State the corollary applied.

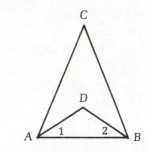

10. In the adjacent figures we know that
    $\angle RST$ and $\angle MON$ are right angles
    $\angle a = \angle b$
    Therefore $\angle c = \angle d$
    State the corollary applied.

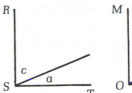

11. In the adjacent figure we know that
    $\angle 1 = \angle 4$
    $\angle 3 + \angle 4 = 180°$
    $\angle 1 + \angle 2 = 180°$
    therefore $\angle 3 = \angle 2$
    State one corollary applied.

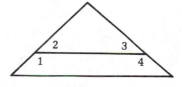

12. In the adjacent figures we know
    $\angle A + \angle B + \angle C = 180°$
    $\angle D + \angle E + \angle F = 180°$
    $\angle A = \angle D$
    $\angle B = \angle E$
    therefore $\angle C = \angle F$
    State two reasons applied.

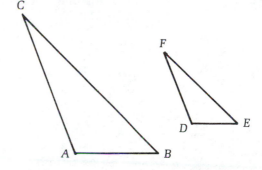

13. $7X - 5 = 30$
    $7X = 35$
    $X = 5$
    State the two axioms applied.

14. $6X - 4X = 20$
    $2X = 20$
    $X = 10$
    State the two axioms applied.

15. $AB = AC$
    $AD = AE$
    $\therefore BD = CE$
    State the axiom applied.

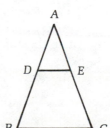

16. $X^4 = 2401$
    $X = 7$
    State the axiom applied.

17. $\angle A = \angle B$
    Both are trisected. (Don't ask how.)
    $\therefore \angle 1 = \angle 2$
    State the axiom applied.

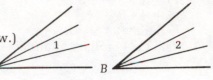

18. $\angle A = \angle B$
    $\angle B = 37°$
    $\therefore \angle A = 37°$
    State the axiom applied.

19. $\angle QAR = \angle RBP$
    $\angle QAR$ is a right angle
    $\therefore \angle RBP$ is a right angle
    State the axiom applied.

20. In $2X^2 + 5X - 6 = 0$   $a = 2, b = 5, c = -6$ and
    $$X = \frac{-b \pm \sqrt{b^2 - 4ac}}{2a}$$
    $$\therefore X = \frac{-5 \pm \sqrt{25 + 48}}{4}$$
    State the one axiom applied.

21. Line segments $AB$ and $BC$ have been bisected at $D$ and $E$ and
    segments $BD$ and $CE$ have been bisected at $F$ and $G$
    $BF = CG$
    $\therefore AB = BC$
    State the one axiom applied.

22. $Y = 2X^2 + 3X - 7$
    $X = -3$
    $\therefore Y = 2$
    State the axiom applied.

23. $AB = PQ$
    $BC = QR$
    $\therefore AC = PR$
    State the axiom applied.

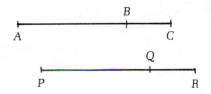

24. Area of rectangle $ABCD = I + II$
    Area of square $AEFD = I$
    $\therefore$Area of the rectangle is greater than area of the square.
    State the corollary applied.

The postulates may be a little more interesting than the axioms, as they pertain to geometric topics, and may be new to you. Remember that these are statements that are accepted without proof, and as such should be "obvious" and readily acceptable by everyone. No counter-examples should exist, or even be imaginable. However, the "truth" of postulates depends only on their consistency. What is "truth" anyway? Each of us has some sort of feeling about what "truth" means. Truth, for the physicist, is *that which works,* and this attitude is characteristic of science. Truth, in practical affairs, usually refers to *whether some event actually took place.* Here, we often take the word of others whom we trust. Truth in spiritual matters, depends upon *faith.* In general, agreement among people regarding spiritual beliefs is difficult to obtain, for there is no test by experience that we can all agree upon.

When we say that a set of mathematical postulates is "true," we mean only **these postulates are free from contradiction.** That is, it is impossible to prove from our postulates that some statement is true and also prove that it is false. Remember that in geometry we do not assert that anything actually exists in the physical world. Our statements have the form: *If* a set of things has certain properties, *then* such-and-such will be true.

Consider the following two statements and determine if they could be "true." That is, are they consistent and non-contradictory?

**In a certain small town there is but one barber.**
**This barber shaves everyone in the town who does not shave himself, and no one else.**
**(Assume that the barber is a man and that all men shave.)**

These two statements would not be acceptable as postulates since they create a contradiction: Who shaves the barber?

The linear accelerator at Stanford is a pretty good example of "straight". The electron beam varies less than $\frac{1}{5}$ inch in a distance of 2 miles. *Stanford Linear Accelerator Center.*

Now to consider our postulates.

*1. One and only one straight line can be drawn through two given points. (Two points determine a line.)*

Since all definitions must use words, and those words in turn need to be defined in terms of other words, if we are to avoid circular definitions, some terms and concepts must remain undefined.

Have you ever had the experience of looking up a word in the dictionary, finding a synonym, which you also did not know, looking it up, and so on, ending with the original word? For instance, if you look up "ductile" you might find it defined as "tractable," look up "tractable" and you find "malleable," look up "malleable" and you find "pliable," and looking up "pliable" you find "ductile!" Of course a good dictionary will give enough synonyms that you should know one of them and not get involved in circular reasoning. But notice that since words are defined only in terms of other words, we must start somewhere with some **undefined** terms.

Point, line, straight, and plane will be undefined terms. It should be realized that we are using the geometric concept of point, in that it has no dimension. A line has no width or thickness, only one dimension of length. A plane has no thickness, only two dimensions of width and length. They are concepts of location and length and surface, and can not actually be drawn. It is interesting that "straight" has its origins in the word for stretched string, and "line" has its origins in the word for linen. Our physical standard for "straight" is the path of light. (The fact that light is "bent" by massive bodies, such as the sun, leads to interesting concepts like space warp.)

*2. Only one plane can be passed through three non-collinear points. (Three points determine a plane.)*

Since collinear points are points which all lie on the same line, non-collinear means not lying on the same line.

It is this concept that makes a tripod or a three-legged stool steady, while a four-legged chair can wobble. Three points will always determine a plane and be steady, while a fourth point may or may not be in the same plane with the other three (co-planar).

**FRANK AND ERNEST**                                              by Bob Thaves

*Reprinted by permission of NEA.*

If leg *A* of the chair is short, or the floor is uneven, the chair will rock back and forth from the plane determined by points *ABC* to the plane determined by the points *BCD*.

Notice that if three points are collinear, there is an infinity of planes containing them. As an example, consider three points on the spine of a book—each of the pages of the book could represent a plane passing through those same three points.

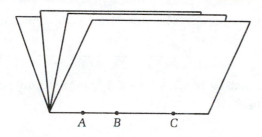

3. *To every pair of points there corresponds a unique positive number. (Distance Postulate)*

Draw the line that passes through a pair of points (this can be done, see Postulate 1). Now coordinate this line with a set of real numbers. Use inches, centimeters, yards, kilometers, or any other unit measure, named or unnamed. $|A - B|$ will now give you the unique (for your particular number line) measure of the distance between the two points. No matter where you place the origin of the number line this distance will always be the same.

| A | B | |
|---|---|---|
| 0 | 7 | $|A - B| = |0 - 7| = 7$ |
| 3 | 10 | $|3 - 10| = 7$ |
| −4 | 3 | $|-4 - 3| = 7$ |
| $-19\frac{1}{4}$ | $-12\frac{1}{4}$ | $|(-19\frac{1}{4}) - (-12\frac{1}{4})| = 7$ |
| 6.39 | 13.39 | $|6.39 - 13.39| = 7$ |

*4. To every angle there corresponds a unique positive number. (Angle Measure Postulate)*

This works in much the same manner as the previous postulate, except here we first need to define what an angle is, and how it is measured. We will also discuss the classification of angles by size and location.

A **line** extends indefinitely in both directions.

A **ray** has an endpoint and extends indefinitely in one direction.

A **line segment** is a portion of a line having two endpoints.

An **angle** is formed by two rays with a common endpoint, called the **vertex** of the angle.

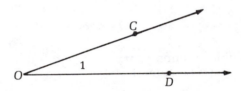

When two lines intersect they form pairs of angles, called **vertical angles,** and some **straight angles.**

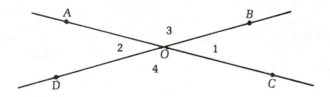

∡1 and ∡2 are vertical angle pairs

∡3 and ∡4 are vertical angle pairs

∡1 and ∡3, ∡2 and ∡3, ∡1 and ∡4, and ∡2 and ∡4 are pairs of angles that form straight angles.

A line can be named with a single small letter: line *a*, or by two of the points on the line: line *AB*. A ray should be named by the vertex and one other point: ray *AB*. An angle can be named in several ways: by the letter at the vertex, $\angle O$, but only if it is the only angle using that vertex; by a number placed inside the angle, $\angle 1$; or by the use of three letters—a point on one ray, the vertex, and a point on the other ray, $\angle COD$ or $\angle DOC$. In the above figure $\angle 1$ and $\angle 2$ are vertical angles, $\angle 3$ and $\angle 4$ are vertical angles, $\angle AOC$ and $\angle BOD$ are straight angles.

Now back to our Angle Measure Postulate. The measure of an angle is a measure of rotation. Imagine two rays, one directly on top of the other.

Now, as ray $AC$ is rotated away from ray $AB$, the measure of the angle is increased.

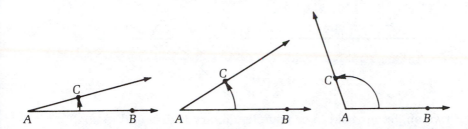

The point $C$ in its rotation all the way around and back to its original position will trace out a circle. We now fit a real number line with exactly 360 units on this full circle and use those units to measure the rotation, or size, of the angle. There are several reasons 360 has been used since ancient times. The early mathematicians were often astronomers also, and the number 360 coincided well with the annual rotation of the earth. Also 360 is a number that is easily divided by many other numbers— making it easy to find fractions of a circle. It is interesting to note that this is one standard of measure that is not metric (powers of 10) that the scientists of the world have agreed to maintain. Each of these 360 units is called a degree, $1°$, and each degree is further subdivided into 60 parts called minutes, $1° = 60'$. Each minute is also subdivided into 60 parts called seconds, $1' = 60''$. Therefore, $1° = 3600''$, which is certainly a very precise measurement! If a right triangle were drawn with an angle of $1''$ and a side 1 km long, the side opposite the angle would be only 5 cm long.

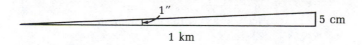

Or if a triangle were drawn from here to the moon with an angle of $1''$, the side opposite the $1''$ angle would be about 2 km long.

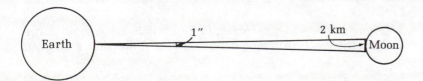

Angles are classified as to size in the following manner:

**Acute:** between 0° and 90°
**Right:** exactly 90°
**Obtuse:** between 90° and 180°
**Straight:** exactly 180°
**Reflex:** greater than 180°

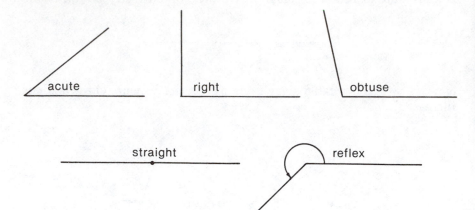

Two angles are said to be **complementary** if they add to 90°.

Two angles are said to be **supplementary** if they add to 180°.

To find the *complement* of an angle we subtract the angle from 90°. The complement of 35° is 90 − 35 = 55°.

To find the *supplement* of an angle we subtract the angle from 180°. The supplement of 112° is 180 − 112 = 68°.

When we wish to find the complement or supplement of an angle measured in minutes and seconds there is a convenient trick we can use. By doing some "borrowing", 90° = 89° 59′ 60″. Likewise, 180° = 179° 59′ 60″. Using these substitute values will greatly simplify the subtraction.

**Find the complement of 23° 31′ 43″**

$$
\begin{array}{r}
89'\ 59'\ 60'' \\
-\ 23°\ 31'\ 43'' \\
\hline
66°\ 28'\ 17'' \text{ is the complement.}
\end{array}
$$

**Find the supplement of 97° 51′ 37″**

$$
\begin{array}{r}
179°\ 59'\ 60'' \\
-\ \ \ 97°\ 51'\ 37'' \\
\hline
82°\ \ \ 8'\ 23'' \text{ is the supplement.}
\end{array}
$$

There are three more postulates which we will list here for completeness' sake, but we will save discussion of them until they are needed:

> 5. *There is a one-to-one correspondence between a set of points and their reflection about a line. (Point Reflection Postulate)*

6. *Through a point not on a line, only one parallel to the line can be drawn. (Euclidean Parallel Postulate)*

7. *Any property of regular polygons which does not depend on the number of sides of the polygon is also true of circles. (Limit Postulate)*

For purposes of further discussion it will be useful to classify triangles. We classify them by characteristics both of angles and legs. By angles, a *triangle* is:

**Acute** if all three of its angles are acute.
**Right** if it contains a right angle. (Only one right angle is possible.)
**Obtuse** if it contains an obtuse angle. (Only one obtuse angle is possible.)
**Equiangular** if all of its angles are equal.

Classified by legs, a triangle is:

**Scalene** if all three legs are of different lengths.
**Isosceles** if two of its legs are equal.
**Equilateral** if all three of its legs are the same length.

Adjectives from both lists may be used in the description of a triangle, such as "obtuse scalene triangle," or "isosceles right triangle," or "right scalene triangle." The student should be able to sketch or visualize the triangles so named. Some, however, are impossible, such as an "equilateral right triangle."

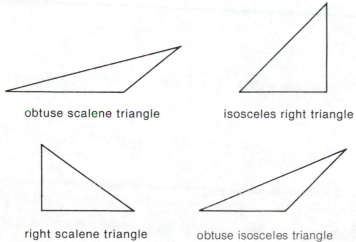

obtuse scalene triangle          isosceles right triangle

right scalene triangle       obtuse isosceles triangle

As mentioned previously, two angles are **complementary** if they form a right angle or the sum of their measures is 90°.

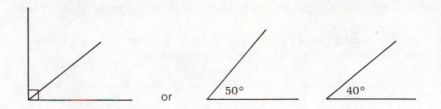

Two angles are **supplementary** if they form a straight angle or the sum of their measures is 180°.

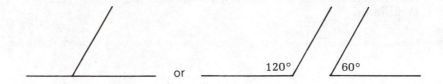

Two angles are **adjacent** if they have the same vertex and a common ray *between* them.

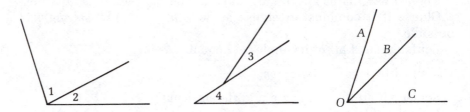

∡1 and ∡2 are adjacent, but ∡3 and ∡4 are not (no common vertex), and ∡AOB and ∡AOC are not (no common side *between* them).

Two lines which intersect to form a 90° angle, or a right angle, are said to be **perpendicular.** (symbol ⊥)

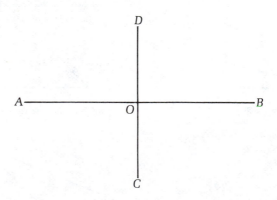

If ∡AOD = 90° then AB ⊥ CD.

We often indicate perpendicular lines with a little square at the vertex of the right angle.

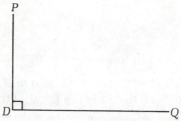

If ⦨*PDQ* is a right angle, then ray *DP* ⊥ ray *DQ*.

In practice we will often use the terms "perpendicular lines," "lines forming a right angle," and "lines forming a 90° angle" interchangeably—since they all describe the same situation.

*AB*, *CD*, *EF*, and *GH* are straight lines concurrent at *O*, *AB* is perpendicular to *CD*.

**EXERCISE SET 5**

1–20. True or false (Mark any statement "false" that is not *certainly* true, from the information given.):

1. ⦨*COB* is a right angle.

2. ⦨*AOC* and ⦨*AOD* are equal adjacent angles.

3. ⦨*AOD* and ⦨*DOF* are adjacent angles.

4. ⦨*AOD* and ⦨*X* are adjacent angles.

5. ⦨*EOC* is half of ⦨*AOC*.

6. ⦨*EOC* is an obtuse angle.

7. ⦨*EOF* is greater than ⦨*COD*.

8. ⦨*HOD* is an acute angle.

9. ⦨*AOD* is supplementary to ⦨*BOC*.

10. ⦨*BOG* is complementary to ⦨*X*.

11. ⦨*BOD* is half as large as ⦨*EOF*.

12. ⦨*BOF* and ⦨*AOH* are vertical angles.

13. ⦨*COH* is greater than ⦨*COG*.

14. ⦨*GOH* is a straight angle.

15. ⦨*AOB* is twice as large as ⦨*AOD*.

16. ⦨*Y* and ⦨*COG* are the same angle.

17. ⦨*X* and ⦨*DHO* are the same angle.

18. ⦨*EOB* is an acute angle.

19. ⦨*COB* and ⦨*COA* are equal.

20. ⦨*COH* and ⦨*DOG* are vertical angles.

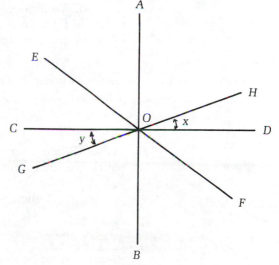

Figure for problems 1–26.

21–25. If ⦨*X* = 23° 34′ 47″ and ⦨*GOE* = 53° 25′ 33″ (assume that pairs of vertical angles are equal), compute:

21. ⦨*DOF* =

22. $\sphericalangle EOH =$

23. $\sphericalangle BOF =$

24. $\sphericalangle AOG =$

25. $\sphericalangle DOE =$

26. Name all the right angles in the figure.

27. Given three distinct lines, $l$, $m$, and $a$, if $a$ is parallel to $l$, and $l$ and $m$ both pass through $P$ then $a$ is not parallel to $m$ because _____

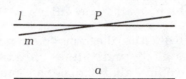

*28. Given that $l$ and $m$ both pass through $P$, $l$ is parallel to $a$, $m$ is parallel to $a$, then _____

29. To every angle there corresponds a unique positive number. In calculus, angles are usually measured in radians instead of degrees. (A radian is the angle formed when the length of arc is equal to the circle's radius.) A complete circumference is $2\pi$ radians and also 360°, and since there is a unique number for any angle, these two measures are equal. Use this information to find a conversion factor for radians; that is,

$$1 \text{ rad} = \underline{\hspace{1cm}}° \quad \text{and} \quad 1° = \underline{\hspace{1cm}} \text{ rad}$$

30. Two points determine a single line. How many lines are determined by 3 non-collinear (not on the same line) points? How many lines are determined by 4 points? By $n$ points?

*31. Three points determine a plane. How many different planes are determined by 4 non-coplanar points? By 5 non-coplanar points? By $n$ points?

32. To every pair of points there corresponds a unique positive number. Measure the distance between $A$ and $B$ in centimeters and in inches. Since both names represent the same number, they must be equal. Use this information to determine a conversion factor; that is,

$$1 \text{ inch} = \underline{\hspace{1cm}} \text{cm} \quad \text{and} \quad 1 \text{ cm} = \underline{\hspace{1cm}} \text{in.}$$

$A \;\longmapsto\!\!\!\!\!\!\!\!\!\!\!\!\!\!\!\!\!\!\!\!\!\!\!\!\!\!\!\!\!\!\!\longmapsto\; B$

---

*"Geometry is the art of right reasoning with wrong figures. The figure on the blackboard is always wrong. The right figure is in your head."—G. Polya*

## 6. Constructions

The basic rules for the construction of geometric figures allow only the use of an unmarked straightedge and a compass. A ruler may be used as the "unmarked straightedge," of course, but measurements are used only when the length of a line segment is specified. Measurements are not used in the process of construction.

In more recent times we have improved upon the methods of Euclid and his friends, as when, in 1553, Benedetti showed that all the ordinary constructions of geometry can be performed with only a straightedge and a compass with a single fixed setting, and in 1797 Mascheroni showed that they can be done with compasses alone.

There are only a few constructions that are necessary, and they will be presented here in a "how-to" manner. After we have discussed formal proofs, and have a few basic theorems at our command, we will prove that each of these construction methods actually does what we claim.

1. To *copy a line segment,* simply take its measure with the compass, place the point of the compass on the line to which you wish to transfer the line segment, and swing an arc to determine the other end of the desired line segment. $AB = CD$

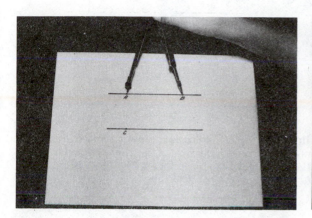
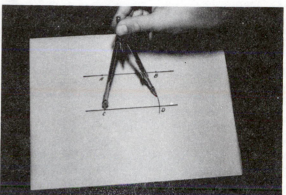

2. To *copy an angle,* since an angle is an amount of rotation, we need an arc upon which to measure. Swing equal arcs across the given angle and one ray of the desired angle. With your compass measure the opening of the given angle and copy that on the desired angle. Draw the other ray through the vertex and the point just located. $\angle AOB = \angle COD$

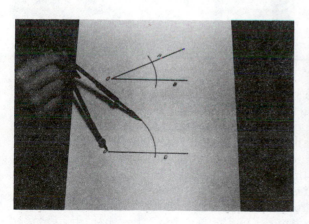
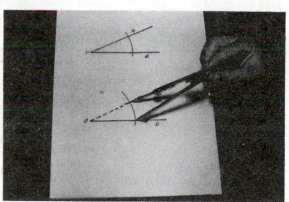

To *bisect* means to divide into two equal parts.

3. To *bisect a line segment,* open your compass to a width greater than half the length of the line segment, swing an arc above and below the line segment from each of its ends, and join the two points just found. This line just drawn will not only bisect the given line segment, but will be perpendicular to it. $AB = BC$

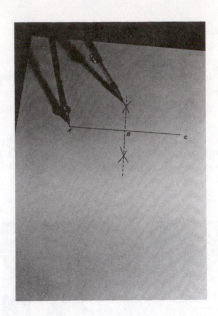

4. To *bisect an angle*, with any setting of the compass swing an arc across the angle. From the arc's intersections with the rays of the angle swing two more arcs in the interior of the angle. Join the point located by these intersecting arcs to the vertex of the given angle, and it will be bisected. $\angle 1 = \angle 2$

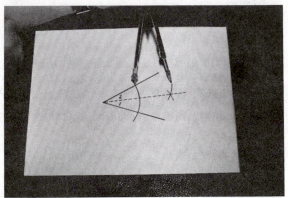

The trisection of an angle is one of the famous problems of geometry. People have been trying to trisect an angle for 2000 years. However, the trisection of a general angle with only straightedge and compass was proven to be impossible by Wantzel in 1837. This was a result of the same work which demonstrated that it was impossible to construct a cube with a volume twice that of a given cube. The National Council of Teachers of Mathematics has published a book on the trisection of an angle: *The Trisection Problem*, by Robert C. Yates. It gives the history of the problem, some mechanical methods using other than straightedge and compass, some famous erroneous attempts at its solution, and some methods which give good *approximations*. The following is one of the simplest of these techniques:

For the arbitrary angle $\angle AOB$ that you wish to trisect, extend $AO$ to $D$, and with $O$ as center draw a semicircle with any convenient radius intersecting $AD$ at $P$ and $Q$.

Find the midpoint, *M*, of arc *PB* by bisecting ∡*AOB*.
Find the midpoint, *N*, of segment *OQ* by bisecting *OQ*.
Draw *MN*.
∡*ANM* will be a very good approximation to one-third ∡*AOB*.

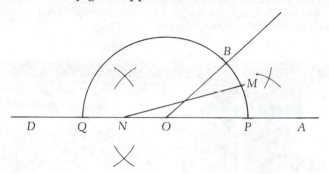

5. **To construct a** *perpendicular to a line at a point on the line*, **with any convenient setting of the compass swing arcs on either side of the given point. Increase the setting of the compass, and from these two points swing arcs above the given point. Join the point located by these arcs to the given point, and the line drawn will be perpendicular.**

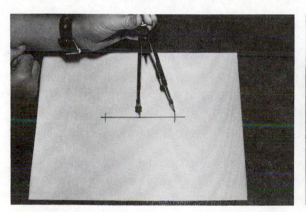
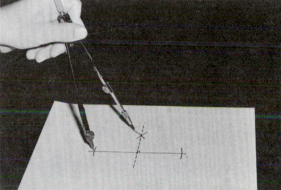

6. **To construct a** *perpendicular to a line from a point off the line*, **set your compass wide enough to swing an arc from the given point that will cut the line in two places. From these two points swing arcs above or below the given line (using any setting of the compass that will allow the arcs to intersect). Join the point just located by the intersection of these arcs to the given point, and it will be perpendicular to the given line.**

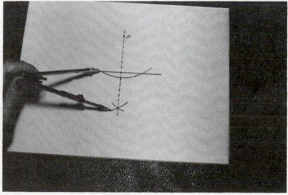

Although "parallel" has not been defined yet, your intuitive concept is sufficient for the next construction.

7. **To construct** *a line parallel to a given line through a given point,* **draw any line through the point that will intersect the given line. Copy the angle formed, putting its vertex at the given point. The extended ray of the new angle will be parallel to the given line.**

 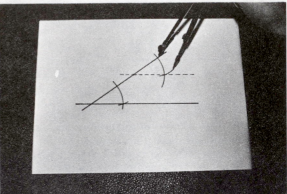

**A POINT OF INTEREST**

Not all objects that can be drawn on paper can be constructed in the form of a three-dimensional model. M. C. Escher enjoyed drawing these kinds of objects. Here is a copy of his lithograph *Ascending and Descending:*

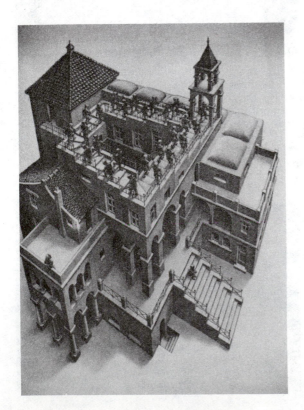

Do the stairs go up, or down? *"Ascending and Descending"* by *M. C. Escher; Escher Foundation—Haags Gemeentemuseum—The Hague.*

If this mill could be constructed we would have perpetual motion for the generation of energy.

Is this an example of perpetual motion? *"Waterfall" by M. C. Escher; Escher Foundation—Haags Gemeentemuseum—The Hague.*

Speaking of energy, this hinge is a modern development (not yet patented) called a sight-activated hinge. It can be opened or closed just by looking at the bottom or top!

A similar invention (also not yet patented) is this self-emptying ashtray.

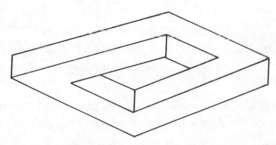

(Use lots of paper and, for these exercises, a ruler.)

1. Construct a 12 cm line segment.
   Construct its perpendicular bisector.
   With the point of intersection of the two lines as center, construct a circle with radius of 4 cm.
   Bisect one of the central angles.
   Construct a perpendicular to the bisector at its intersection with the circle.

2. Use a ruler to construct a right triangle with legs of 5 cm and 12 cm.
   What is the measure of the hypotenuse? (The hypotenuse is the side opposite the right angle.)

3. Construct an equilateral triangle with sides of 10 cm.
   Construct a line through one vertex parallel to the opposite side.
   With your protractor, find the measure of the angle formed by this parallel line and the side adjacent to it.

4. Construct any triangle.
   Without measuring any sides or angles, construct a second triangle the same size and shape.

5. Without measuring, construct an angle of $22\frac{1}{2}°$. Hint: $\frac{1}{2}(45°) = 22\frac{1}{2}°$

6. Without measuring, construct an angle of 15°. Hint: Start with an equilateral triangle. (Each angle is 60°.)

7. Construct an obtuse scalene triangle.
   Construct the perpendicular bisectors of all three sides and produce them until they intersect.

8. Construct an acute isosceles triangle.
   Bisect one of the legs. Call the point of bisection $P$.
   Through $P$ construct a line parallel to the base.
   Does this parallel seem to intersect the other leg at its midpoint?

9. Draw any angle. Now, construct a second angle with its sides parallel to the sides of the first angle.

10. Draw any angle. Now, construct a second angle with its sides perpendicular to the sides of the first angle.

11. Draw any four-sided figure. Construct the midpoints of each side and join the opposite midpoints. What property do these lines seem to have?

12. Construct a parallelogram. (A four-sided figure with its opposite sides parallel.)

13. Construct an isosceles trapezoid. (A four-sided figure with one pair of sides parallel and the other pair of sides equal but not parallel.)

14. Draw two line segments. Label one $a$ and the other $b$. Now construct a line segment of length $2a + 3b$.

15. Draw any two angles. Label one $\angle A$ and the other $\angle B$. Now construct an angle which measures $2\angle A + \angle B$.

16. Draw any angle less than 45°, call it $\angle B$. Now construct a triangle $ABC$ such that $\angle A = 3 \angle B$.

17. Divide a 7 cm line into four equal parts by construction.

18. Construct a 45° angle. From any point on one side of the angle construct a perpendicular to the other side. Compare the length of this perpendicular to its distance from the vertex of the angle.

19. Draw any line segment. Construct a square with that length as a side.

20. Draw any line segment. Construct an equilateral triangle with that length as a side.

21. Construct an acute scalene triangle. Construct the perpendiculars from each vertex to the opposite side.

22. Construct a scalene right triangle. Construct the bisectors of each angle.

23. Construct triangle $ABC$ where $\angle A = 60°$, $\angle B = 45°$, and $AB = 6$ cm.

24. Construct the triangle $ABC$ where $\angle A = 45°$, $AB = 6$ cm, and $AC = 7$ cm.

25. Construct triangle $ABC$ where $AB = 5$ cm, $BC = 6$ cm, and $AC = 7$ cm.

*26. Try creating a non-constructible model like those on the previous pages. If you have difficulty, you might begin with a triangular self-emptying ashtray.

## 7. Transformations

One of the basic assumptions used by Euclid was that a figure could be moved without changing its size or shape. In proving some fundamental theorems he used the concept of superposition—that is, the placing of one figure on another to see if they would coincide exactly, and therefore be the same size and shape. In the figures below, the triangle $A'B'C'$ would be moved on top of triangle $ABC$ in such a manner that $A$ would coincide with $A'$, $B$ with $B'$, and $C$ with $C'$.

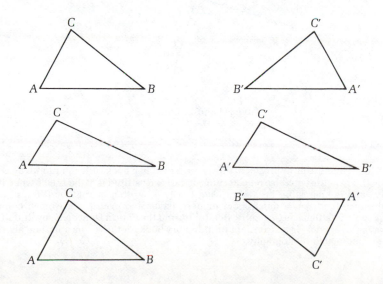

While this moving of a geometric figure may be intuitively acceptable, there are some difficulties here. Notice that in order to make the first pair of triangles coincide, $A'B'C'$ needs to be "turned over." That is, it must be lifted out of the plane, turned over, and then returned to the plane. This requires going from the two-dimensional space into three-dimensional space. What will we do if we wish to make three-dimensional solids coincide? We would have to lift them from three-dimensional space into the fourth dimension, turn them over, and return them to the third dimension. While this excursion into the fourth dimension should not bother a mathematician, it is very difficult to visualize.[2]

This business of showing that figures are identical can be put on a firmer mathematical basis by considering one figure as a transformation of the other. A **transformation** is a one-to-one correspondence between two sets of points in the plane. If we can find an isometric (distance preserving) transformation that will map $A$ into $A'$, $B$ into $B'$, and $C$ into $C'$, then the figures will be identical. The transformation approach to geometry has had a renewal of interest in recent years, and some texts develop the entire geometry with the use of transformations. We do not intend to do that—we present the concepts of transformations only in order to establish a few of the basic theorems, and then we will proceed in the more traditional manner.

Our main objective in this section is to establish the *existence* of a transformation that will move one figure onto another through a legitimate algebraic process. With this tool we will be able to *prove* some of the theorems in later sections that most texts postulate. (Especially the triangle congruency theorems SAS, ASA, and SSS.)

There are three types of transformations we wish to consider: reflection, translation, and rotation. (These were illustrated by the first, second, and third sets of triangles given at the beginning of this section.) We will be using some of the construction methods learned in the last section.

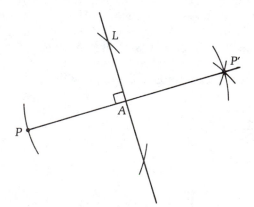

Given P and L, finding P'.

[2]A light, interesting approach to this is presented in a book called *Flatland*, by Edwin A. Abbott. This little book also reflects some of the sexist attitudes of the 1880's when it was written, and is interesting in that respect. A sequel, written in 1965 by Dionys Burger, called *Sphereland*, has a much more modern flavor and even gives women credit for having some intelligence! *Geometry, Relativity, and the Fourth Dimension* by Rudolf V. B. Rucker, a Dover book, is an excellent introductory book for these concepts, and also has a very fine annotated bibliography.

**Reflection:** *P'* is a reflection of *P* through line *L*, if and only if *L* is the perpendicular bisector of the line *PP'*. (A point on *L* is its own reflection.) In the top figure on page 53, *A' B' C'* is a reflection of *ABC*.

*L* is called the **axis of symmetry,** the axis, or the mirror. When given a point *P* and an axis *L*, we can construct *P'* by constructing the perpendicular from *P* to *L*, call the point of intersection *A*, and copying *PA* on the extended perpendicular.

Once you have some theorems at your disposal, you will be able to prove that *P'* will always be at the intersection of your construction marks for the perpendicular if you do not change your compass settings.

If two points are given, *P* and *P'*, one can be demonstrated to be the reflection of the other by constructing the perpendicular bisector of *PP'*—this will be the required axis, *L*.

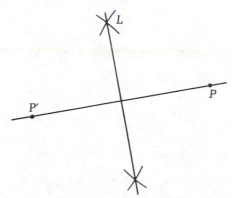

Given P and P', finding L.

If we wish to reflect a triangle *ABC* about the axis *L*, we reflect each of the vertices about *L* to locate *A'*, *B'*, and *C'*, and then join them to form the reflected triangle *A' B' C'*. Since two points determine a line segment, say *A* and *B*, and their reflections, *A'* and *B'* also determine a line segment, line segment *A'B'* is the reflection of line segment *AB*. The two line segments are two sets of points with a one-to-one correspondence (by the Point Reflection Postulate).

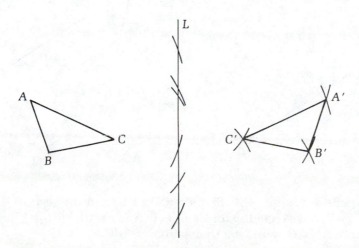

If we are presented with two triangles, *ABC* and *PQR*, and wish to determine if one is the reflection of the other, use one pair of corresponding points, say *A* and *P*, to construct the axis *L*, then reflect *B* and *C* about *L* and see if they coincide with *Q* and *R*. If *B'* coincides with *Q*, and *C'* coincides with *R*, then one triangle is the reflection of the other.

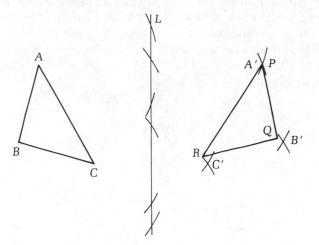

In this case, △ *PQR* is not a reflection of △ *ABC*, since *C'* does not coincide with *R* and *B'* does not coincide with *Q*. You should realize that this technique cannot be used to "prove" a figure is a reflection, since it depends upon the accuracy of your construction.

The other two transformations, the translation and the rotation, can be defined in terms of the reflection. The **translation** of a figure is two consecutive reflections through parallel lines. It is easy to think of as a sliding motion in a horizontal and vertical direction. *E'F'G'* is a translation of *EFG*.

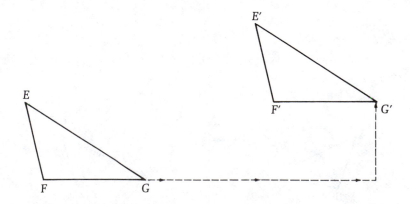

Let us reflect triangle *ABC* first about axis $L_1$ and the resulting figure *A'B'C'* will be reflected again about $L_2$ (a line parallel to $L_1$). The resulting figure *A" B" C"* will be a translation of *ABC*.

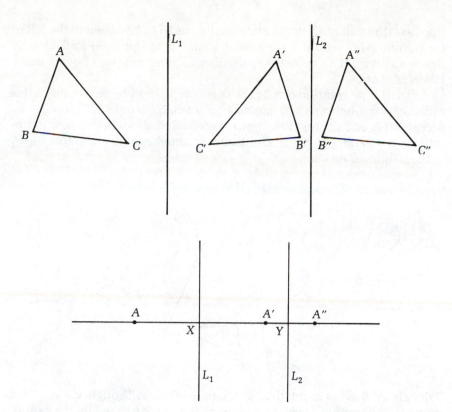

Consider now the translation of $A$ about $L_1$ and $L_2$ to $A''$. The distance which it is translated is the line segment $AA''$, but $AA'' = AX + XA' + A'Y + YA''$ and since $L_1$ is the perpendicular bisector of $AA'$, $AX = XA'$. By the same reason $A'Y = YA''$, so if we substitute into the first equation we get $AA'' = XA' + XA' + A'Y + A'Y$ or $AA'' = 2(XA') + 2(A'Y)$ by distributing $AA' = 2(XA' + A'Y)$. However, since $XA' + A'Y = XY$ we have that $AA'' = 2XY$. That is, the distance a figure is translated will always be twice the distance between the parallel lines.

If we are given a figure and are asked to translate it a given distance in a given direction, draw axes $L_1$ and $L_2$ perpendicular to the given direction and half the given distance apart. Their distance from the given figure is immaterial. Example: Translate the triangle $ABC$ 6 cm in the direction indicated.

First construct $L_1$ perpendicular to the desired direction (it can be put anywhere). Then construct $L_2$ half the desired distance ($XY = 3$ cm) in

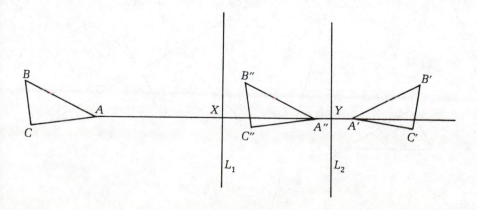

the direction you wish to translate the figure. (If $L_2$ had been to the left of $L_1$ triangle $ABC$ would have been translated to the left.) Reflect your figure about $L_1$, then reflect it again about $L_2$ and you will have a translation of 6 cm.

Finally the **rotation** of a figure is accomplished by two consecutive reflections through a pair of intersecting lines. If triangle $ABC$ is reflected first about $L_1$ and the resulting figure is reflected about $L_2$ this new figure will be a rotation of $ABC$. The point of intersection of $L_1$ and $L_2$, $O$, is the center of rotation.

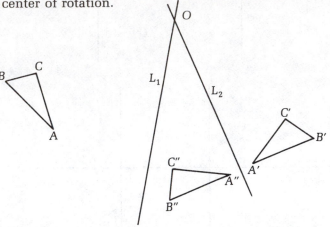

Triangle $A''B''C''$ is a rotation of triangle $ABC$. Although we will not prove it, the amount of rotation is twice the angle formed by the axes $L_1$ and $L_2$.

To determine if $\triangle PQR$ is a rotation of $\triangle ABC$ we proceed as follows:

Join two pairs of corresponding points, and construct the perpendicular bisectors to locate $O$, the center of rotation.

Draw any axis $L_1$ between the two figures, through $O$.

Reflect $\triangle ABC$ about $L_1$ to form $\triangle A'B'C'$.

Join a pair of corresponding vertices of $\triangle A'B'C'$ and $\triangle PQR$. ($A'$ and $P$ in this illustration)

Construct the perpendicular bisector of this line—it is $L_2$.

Reflect the remaining points of $\triangle A'B'C'$ about $L_2$.

If these reflected points coincide with their corresponding points in $\triangle PQR$, then $\triangle PQR$ is a rotation of $\triangle ABC$.

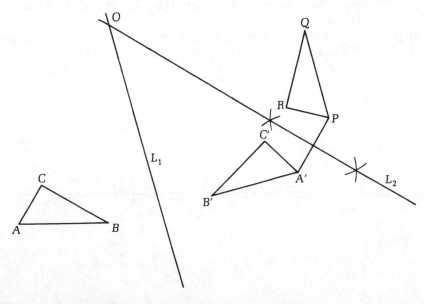

There are several theorems concerning these transformations that we will state, but will not attempt to prove. The student should realize, however, that these are theorems, and not postulates, and as such can be proven.

When any of the above transformations are performed, that is, reflection, translation, or rotation, the following hold true:

1. The distances between points in the original figure and the corresponding points in the transformed figure are equal.
2. The property of betweeness is preserved. That is, if $P$ is between $A$ and $B$ in the original figure then $P'$ will be between $A'$ and $B'$ in the transformed figure.
3. Angle measure is preserved. That is, $\angle A'B'C'$ will have exactly the same measure as $\angle ABC$.
4. Collinearity is preserved. That is, if points $P$, $Q$, and $R$ lie on a straight line in the original figure, then $P'$, $Q'$, and $R'$ will lie on a straight line in the transformed figure.

1. Construct the reflection of each of the following points about the given line. Label them $A'$, $B'$, and $C'$.

**EXERCISE SET 7**

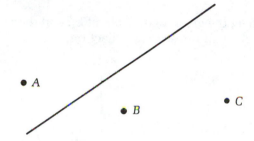

2–7. Reflect each of the figures about line L. Label with prime notation (').

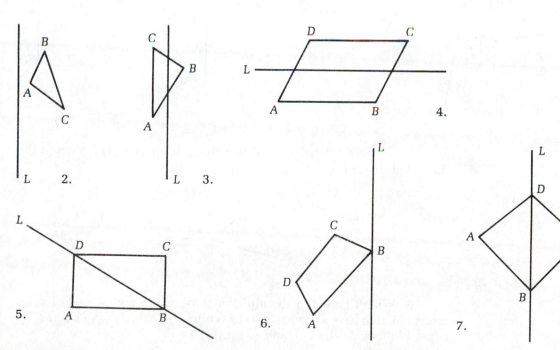

8–12. Sketch *all* axes of symmetry for each of the following figures.

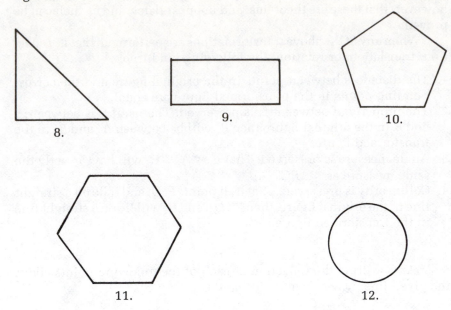

8.

9.

10.

11.

12.

13–14. Reflect the given figures about $L_1$, and then $L_2$. Call the first reflection of $A$, $A'$, and the second reflection, $A''$.

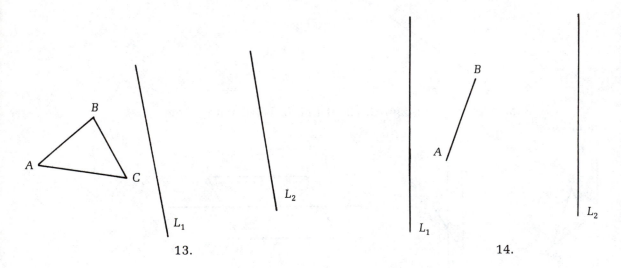

13.

14.

15. Perform a translation so that the triangle is moved 6 cm in the indicated direction.

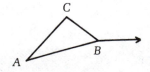

16. Which letters of the alphabet have a horizontal axis of symmetry? Which have a vertical axis of symmetry? Which have both? (Use capital letters; A, B, C, D, and so on.)

17. Make a word that has a vertical axis of symmetry.

18. Make a word that has a horizontal axis of symmetry.

19–21. Rotate the following figures by first reflecting about $L_1$ and then about $L_2$. Call the first reflection of $A$, $A'$, and the second reflection, $A''$.

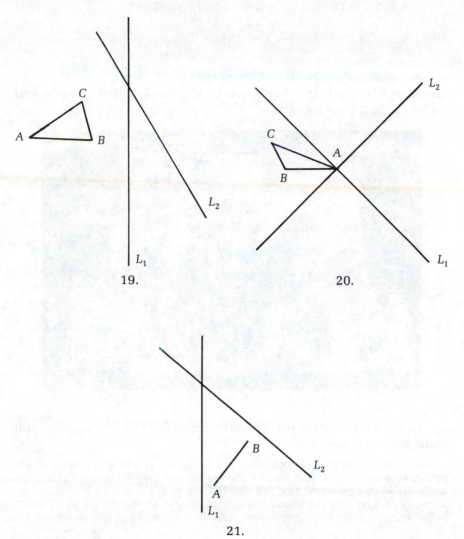

19.                                          20.

21.

22. Find lines $L_1$ and $L_2$ that will rotate the triangle $ABC$ into the position of triangle $A'B'C'$.

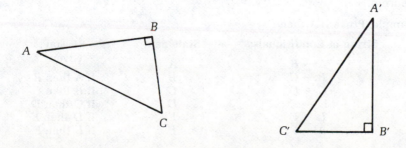

23. What are the next three symbols in this sequence?

$$\text{Ⅱ} , \text{♡} , \text{Ɛ3} , \text{♇4} , \text{Ɔ5} , \text{ȝ6} , \underline{\quad} , \underline{\quad} , \underline{\quad}$$

24. What is the next number in this sequence?

9, 18, 36, 72, 144, \_\_\_\_\_

Hint: It is unrefined, coarse, and vulgar. It is too gross!

25. In a deck of cards, which of the hearts have a vertical axis of symmetry? Which have a horizontal axis? (Disregard the numbers and J, Q, K, A in the corners.)

## 8. Formal proof

G. H. Hardy (1877–1947), a prominent English philosopher and mathematician, has said:

A mathematician, like a painter or a poet, is a maker of patterns. If his patterns are more permanent than theirs, it is because they are made of ideas. The mathematician's patterns, like the painter's or poet's, must be beautiful; the ideas, like the color or the words, must fit together in a harmonious way. Beauty is the first test; there is no permanent place in the world for ugly mathematics.

The classical formal proof in geometry is very much like a string of conditional statements. The format is changed so that the original hypothesis, as given by the theorem to be proved, is listed first, and then is followed by successive conclusions—until the conclusion of the original theorem to be proved is reached. **Each conclusion in the list is justified by a definition, axiom, postulate, or previously proven theorem.**

**Example: Prove if $A$ then $F$.**

| String of Conditionals: | Statement | Reason |
|---|---|---|
| | | Given |
| $A \rightarrow B$ | $A$ | |
| $B \rightarrow C$ | $B$ | If $A$ then $B$ |
| $C \rightarrow D$ | $C$ | If $B$ then $C$ |
| $D \rightarrow E$ | $D$ | If $C$ then $D$ |
| $E \rightarrow F$ | $E$ | If $D$ then $E$ |
| $\therefore A \rightarrow F$ | $F$ | If $E$ then $F$ |

For us to arrive at valid conclusions, all of the conditional statements used in the right-hand column for justifications must be valid. Which is why we must use only definitions, axioms, postulates, and previously proven theorems.

The format for the classical proof of a theorem usually contains four parts:

1. **A statement of the theorem to be proved.** This step is omitted when we are proving an exercise, where the hypotheses are too specific, and the conclusions usually not general enough, to be of future use.
2. **An accurate figure illustrating the statement.** Be certain the figure is general in nature and does not assume anything not given in the hypothesis.
3. **The statement of the hypothesis and conclusion in specific terms of the figure.** Often these are renamed "Given" and "Prove." Add letters and numbers to the figure as necessary so that all parts can be conveniently referred to.
4. **The body of the proof—which consists of two columns; a list of successive conclusions on the left and the justifying statement on the right.** Occasionally, a paragraph-type proof is given—where, instead of two columns, a series of conditional statements is given. These are valid reasoning, of course, but are usually harder to follow if of any length and are more susceptible to subtle errors.

Now, as an illustration to all of the above, let us look at a proof attributed to Thales. It is an important theorem that we will use often, and yet is very simple to prove.

Many students feel they need a miracle to fill in a proof between the "Given" and the "Prove"!

*Cartoon by S. Harris.*

"I THINK YOU SHOULD BE MORE EXPLICIT HERE IN STEP TWO."

> *If two straight lines intersect, the vertical angles are equal.*

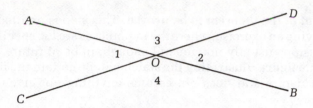

*Given:* Straight lines *AB* and *CD* intersecting at *O*.

*Prove:* ∡1 = ∡2 and ∡3 = ∡4.

| Statements | Reasons |
|---|---|
| **1.** *AB* and *CD* are straight lines | **1.** Given |
| **2.** ∡*AOB* and ∡*COD* are straight angles | **2.** Definition of straight angle |
| **3.** ∡1 and ∡3 are supplementary<br>∡2 and ∡3 are supplementary | **3.** Definition of supplementary |
| **4.** ∡3 = ∡3 | **4.** Reflexive property |
| **5.** ∡1 = ∡2 | **5.** Supplements of equal angles are equal |
| **6.** ∡2 and ∡3 are supplementary<br>∡2 and ∡4 are supplementary | **6.** Definition of supplementary |
| **7.** ∡2 = ∡2 | **7.** Reflexive property |
| **8.** ∡3 = ∡4 | **8.** Supplements of equal angles are equal |

**NOTE:** The last three steps are really not necessary, as the figure was general and the numbering of the angles arbitrary. Either pair could have been called angle 1 and angle 2, so proving one pair equal really proves both pairs are equal.

As another example of a formal proof, consider the theorem:

> *If two straight lines intersect to form equal adjacent angles, then the lines are perpendicular.*

Given:  lines $AB$ and $CD$ intersecting at $O$
       ∡1 = ∡2

Prove:  $AB \perp CD$

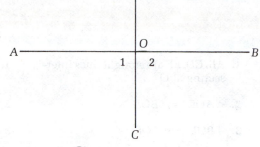

| | |
|---|---|
| 1. $AB$ and $CD$ are straight lines | 1. Given |
| 2. ∡$AOB$ is a straight angle | 2. Definition of straight line |
| 3. ∡$AOB$ = 180° | 3. Definition of straight angle |
| 4. ∡1 + ∡2 = ∡$AOB$ | 4. Whole = sum of parts |
| 5. ∡1 + ∡2 = 180° | 5. Substitution |
| 6. ∡1 = ∡2 | 6. Given |
| 7. 2 ∡1 = 180° | 7. Substitution |
| 8. ∡1 = 90° | 8. Division |
| 9. $AB \perp CD$ | 9. Definition of $\perp$ |

As further examples of the form a proof should take we will do several simple exercises. Since they are exercises and not theorems, the first part of a proof—the statement of the theorem—will be omitted.

Notice that in the proofs of these theorems and exercises each reason used is an axiom, postulate, definition, or previously proven theorem. As such, it is a conditional statement, and before it can be used as a reason, we must demonstrate that its hypothesis has been met.

**Example 1:**

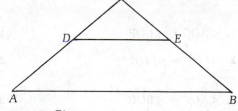

Given: $AC = BC$
      $AD = BE$

Prove: $CD = CE$

| | |
|---|---|
| 1. $AC = BC$ | 1. Given |
| 2. $AD = BE$ | 2. Given |
| 3. $AC - AD = BC - BE$ | 3. Subtraction axiom |
| 4. $CD = CE$ | 4. Substitution |

**Example 2:**

*Given:* Straight lines *AB*, *CD*, and *EF*
intersecting at *O*.
$\angle AOC = \angle BOF$

*Prove:* $\angle BOD = \angle AOE$

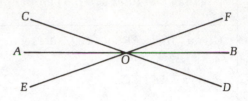

| | |
|---|---|
| **1.** *AB, CD, EF* are straight lines intersecting at *O* | **1.** Given |
| **2.** $\angle AOC = \angle BOD$ | **2.** Vertical angles are = |
| **3.** $\angle BOF = \angle AOE$ | **3.** (2) |
| **4.** $\angle AOC = \angle BOF$ | **4.** Given |
| **5.** $\angle BOD = \angle AOE$ | **5.** Quantities equal to equal quantities are equal to each other |

**Example 3:**

*Given:* Straight lines *AB*, *CD*, and *EF*
intersecting at *O*.
$\angle AOC = \angle BOF$

*Prove:* $\angle AOF = \angle BOC$

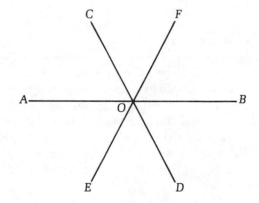

| | |
|---|---|
| **1.** $\angle AOC = \angle BOF$ | **1.** Given |
| **2.** $\angle COF = \angle COF$ | **2.** Reflexive property (sometimes called "Identity") |
| **3.** $\angle AOF = \angle BOC$ | **3.** Addition |

As an aid in the recollection of theorems and corollaries that might be helpful as reasons, Appendix III is a list of Theorems and Corollaries.

Learning to write your own proofs is one of the more difficult tasks of geometry. Skill in writing proofs, like learning to ride a bicycle, paint a picture, or play a good game of tennis, takes practice. It will take some time to learn the format and the necessary degree of rigor.

Use classical format to write a formal proof for each of the following exercises.

1. Given: Straight line $AB$ with $\angle 3 = \angle 4$

   Prove: $\angle 1 = \angle 2$

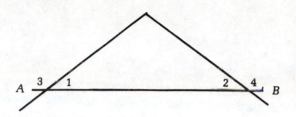

2. Given: Straight line $AB$ with $\angle 3 = \angle 4$

   Prove: $\angle 1 = \angle 2$

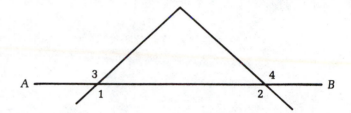

3. Given: $\angle 1 = \angle 3$

   Prove: $\angle 2 = \angle 4$

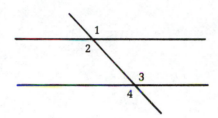

4. Given: $AD = BE$, $DC = EC$
   Prove: $AC = BC$

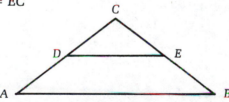

5. Given: $\angle 1$ complementary to $\angle 2$
   $\angle 1 = \angle 3$

   Prove: $\angle 3$ complementary to $\angle 4$

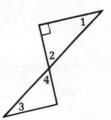

6. Given: Straight lines $AB$, $CD$, and $EF$
   intersecting at $O$
   $\angle AOC = \angle BOF$

   Prove: $\angle AOD = \angle BOE$

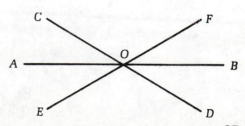

7. Given: $\angle BAD = \angle CAE$

   Prove: $\angle BAC = \angle DAE$

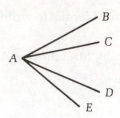

8. Given: Straight lines $MN$ and $XY$
   intersecting at $P$
   $\angle 1 = \angle 2$

   Prove: $\angle 1$ is a right angle.

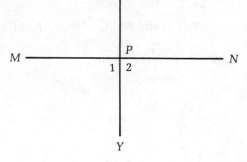

9. Given: $\angle ABC = \angle DEF$
   Both are bisected

   Prove: $\angle 1 = \angle 2$

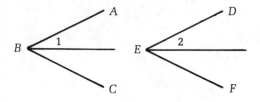

10. Given: $AB + BC = AC$
    $AB = CD$

    Prove: $CD + BC = AC$

11. Given: $\angle DOX + \angle XOG = \angle DOG$

    Prove: $\angle DOG > \angle DOX$

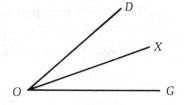

12. Given: $DB = EC$
    $AD = AE$

    Prove: $BA = AC$

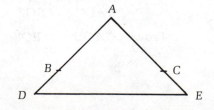

13. Given: ∢1 = ∢2
       ∢2 = ∢3
       ∢3 = ∢4

  Prove: ∢1 = ∢4

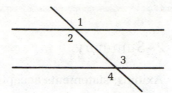

14. Given: ∢RST and ∢MON are right angles
       ∢a = ∢b

  Prove: ∢c = ∢d

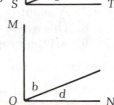

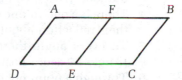

15. Given: ∢A + ∢B + ∢C = 180°
       ∢D + ∢E + ∢F = 180°
       ∢A = ∢D
       ∢B = ∢E

  Prove: ∢C = ∢F

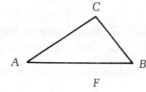

16. Given: $AB = DC$
       $E$ bisects $DC$
       $F$ bisects $AB$

  Prove: $AF = EC$

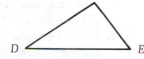

*17. An interesting pastime, which illustrates the property of a given piece of information being transformed in successive steps into a new piece of information, is word chains. For instance, HATE can be transformed into LOVE in the following manner, HATE, HAVE, HIVE, LIVE, LOVE. The rules allow only one letter to be transformed in each step, and all steps must be valid words. Change CAT to DOG.

*18. Change PIER to DOCK.

*19. Change COLD to WARM.

*20. Change HOME to LAND.

*21. Change HEART to SPADE.

*22. See if you can get BLOOD out of STONE.

Notice that there is often more than one route to the same result. The same holds true for geometric proofs.

## 2—Summary

Axioms: statements accepted without proof

1. If equals are added to equals, the sums are equal.
2. If equals are subtracted from equals, the differences are equal.
3. If equals are multiplied by equals, the products are equal.
4. If equals are divided by non-zero equals, the quotients are equal.
5. Like powers of equals are equal.
6. Like positive roots of equals are equal.
7. A quantity may be substituted for its equal in any expression.
8. The whole of a quantity is equal to the sum of its parts.
9. All the properties of the real numbers.

Corollaries to the axioms:

1. Halves of equals are equal.
2. Doubles of equals are equal.
3. Quantities equal to the same or to equal quantities are equal to each other.
4. Equal angles have equal complements.
5. Equal angles have equal supplements.
6. The whole of a quantity is greater than any of its parts.

Postulates: geometric statements accepted without proof

1. One and only one straight line can be drawn through two points. (Two points determine a line.)
2. Only one plane can be passed through three non-collinear points. (Three points determine a plane.)
3. To every pair of points there corresponds a unique positive number. (Distance Postulate.)
4. There is a one-to-one correspondence between a set of points and their reflection about a line. (Point Reflection Postulate.)
5. To every angle there corresponds a unique positive number. (Angle Measure Postulate.)
6. Through a point not on a line only one parallel to the line can be drawn. (Euclidean Parallel Postulate.)
7. Any property of regular polygons which does not depend on the number of sides of the polygon is also true of circles. (Limit Postulate.)

Constructions:

1. Copy a line segment.
2. Copy an angle.
3. Bisect a line segment.
4. Bisect an angle.
5. Construct a perpendicular to a line at a point on the line.
6. Construct a perpendicular to a line from a point off the line.
7. Construct a line parallel to a given line through a given point.

Transformations: a one-to-one correspondence between two sets of points

Reflection: $P'$ is a reflection of $P$ through $L$ if and only if $L$ is the perpendicular bisector of the line $PP'$.

Translation: Two consecutive reflections through parallel lines.
Rotation: Two consecutive reflections through intersecting lines.

The Parts of a Classical Proof:

A statement of the theorem to be proven.
An accurate figure illustrating the statement.
The statement of the hypothesis and conclusion in specific terms of the figure.
The body of the proof—two columns—statements on the left, and justifying reasons on the right.

Theorems:

If two straight lines intersect, the vertical angles are equal.
If two straight lines intersect to form equal adjacent angles, then the lines are perpendicular.

Essentially triangular in shape, the Hyatt Regency's predominant exterior materials are sand-blasted concrete and bronze solar glass set in dark bronze frames. Ceramic tiles with an intersecting circle design pave extensive areas of outside walks and lower interior floors. *Hyatt Regency, San Francisco.*

# CHAPTER 3

# BASIC THEOREMS

In this chapter we will prove the more basic theorems which, in turn, are used so often in the proofs of complex theorems. There are three major theorems about congruent triangles, a couple of theorems about isosceles triangles, and some theorems about parallel lines. We will then use a number of these theorems to prove that the manipulations used in our constructions actually do what we said they would do. We will also introduce the use of auxiliary lines in a proof.

The material in Section 26—Non-Euclidean geometries—will fit in nicely after Section 12—Parallel lines—if time permits.

A pair of triangles are said to be **congruent** if there is a correspondence between the vertices such that the corresponding sides are equal and the corresponding angles are equal. That is, the triangles will

## 9. Congruent triangles—SAS and ASA

Triangles are used extensively in construction due to their strength and rigidity. The curve of the power lines is called a catenary.

have to have all of their corresponding parts equal. The symbol for congruent is ≅, which is composed of two parts—the "=" indicating they are the *same size*, and the "~" indicating they are the *same shape*.

In the figure below, if ∢A = ∢D, ∢B = ∢E, ∢C = ∢F, AB = DE, BC = EF, and AC = DF, then by definition △ABC ≅ △DEF.

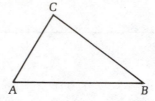 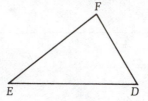

Conversely, if we know the triangles to be congruent, then we know that all of the corresponding parts are equal. This is a lot of information gained when the triangles can be shown to be congruent—six pairs of equalities!

We use this technique so often that we have found it convenient to abbreviate the statement:

> *"Corresponding parts of congruent triangles are equal"* **(abbreviated as CPCTE.)**

The concept of congruent triangles is a very powerful tool. It is one of the most frequently used concepts in elementary geometry. But notice that CPCTE says *if* the triangles are congruent, *then* we have all these convenient equalities. It can never be used as a reason in a proof until we have first established that the triangles are, in fact, congruent. Obviously, the definition is not a very efficient way to proceed, so we need some shortcuts.

Fortunately, there are several convenient ways to prove triangles congruent, but before we look at them we need some terminology. An angle is **included** by the two sides, or rays, which form it. A side is **included** by the two angles which both use it as a side. In triangle ABC, ∢B is included by sides AB and BC. Side BC is included by ∢B and ∢C.

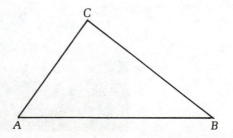

First an intuitive approach: If everyone in the class were asked to construct a triangle with a 7 cm side included by angles of 60° and 40°, would all the triangles be identical? (Ignoring construction errors.) If they were all cut out of the paper, would they stack neatly—that is, would they all be the same size and shape? Some might have to be "flopped over," that is, reflected, but they would all be congruent. It would seem that all the triangles with two angles and the included side equal would be congruent. Now suppose everyone were asked to construct an angle of 50°, mark off 6 cm on one ray and 8 cm on the other ray, and then join these points to form a triangle. Would all these triangles be congruent? If cut out, would they make a nice neat stack? Once again, let's ask our group of experimenters to construct a triangle. This time the three sides are 5 cm, 7 cm, and 9 cm. Will all the triangles be the same size and shape? I hope, in each of the three cases, you were convinced that the given conditions were sufficient to guarantee that all of the triangles constructed would be congruent. Of course, experimenting and arriving at a conclusion in this manner is inductive reasoning, and we would like to prove the results deductively. Stated as theorems:

> *If two triangles have two sides and the included angle of one equal respectively to two sides and the included angle of the other, then they are congruent.* **(abbreviated as SAS.)**
>
> *If two triangles have two angles and the included side of one equal respectively to two angles and the included side of the other, then they are congruent.* **(abbreviated as SAS.)**
>
> *If two triangles have three sides of one equal to three sides of the other, then they are congruent.* **(abbreviated as SSS.)**

Most texts either present these three statements as postulates, or postulate one and then use it to prove the other two. What we would like to do here is to "move" the triangles to demonstrate that the parts given as equal are sufficient to make all corresponding parts equal. We do not want to move them in the manner of Euclid, but to use the concepts of transformations to demonstrate that there is a one-to-one correspondence between the two sets of points which form the triangles. (In Book I of *Elements* Euclid defined congruence as a "Common Notion." "Things which coincide with one another are equal to one another.")

We will write a proof using the transformation concepts for SAS, ASA, and SSS theorems, prove some additional theorems about triangles, and then present some examples illustrating the uses of these theorems.

From our brief work with the transformations, it should be apparent that there will always be a reflection that will transform any point to any other, and there will always be a rotation that will superimpose any line upon another. We will use these concepts in some of the proofs which follow. Keep in mind also that under the three main transformations—reflection, translation, and rotation—distances, angle measures, betweenness, and colinearity remain unchanged.

> *If two sides and the included angle of one triangle are equal
> respectively to two sides and the included angle of another
> triangle then the triangles are congruent.* (SAS)

Given: $AB = DE$, $\angle B = \angle E$, $BC = EF$

Prove: $\triangle ABC \cong \triangle DEF$

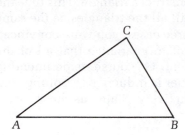
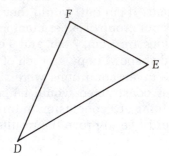

| | |
|---|---|
| 1. Rotate triangle *DEF* so that *E* coincides with *B* and *DE* falls on *AB* | 1. Existence of a rotation |
| 2. $AB = DE$ | 2. Given |
| 3. *D* will coincide with *A* | 3. Distance Postulate (*AB* determines a unique number, call it *X*, since $AB = DE$, $DE = X$, but only one point can have this number, so *D* and *A* coincide.) |
| 4. If *C* and *F* are on opposite sides of *AB*, reflect $\triangle DEF$ about *AB* | 4. Existence of a reflection |
| 5. $\angle E = \angle B$ | 5. Given |
| 6. *EF* will coincide with *BC* | 6. Angle Measure Postulate [used as in (3)] |
| 7. $BC = EF$ | 7. Given |
| 8. *F* will coincide with *C* | 8. (4) |
| 9. *DF* will coincide with *AC* | 9. Two points determine a unique line |
| 10. $\triangle ABC \cong \triangle DEF$ | 10. They coincide throughout which guarantees all their corresponding parts are equal |

Notice, in the proof above, that the numbering of the steps allows us
to use the number of a preceding step as a reason, rather than writing
out the reason a second time. As an example of how this theorem can be
used in proofs, consider the following examples:

*Given:* Quadrilateral *ABCD* with
diagonals intersecting at *E*,
*AE = CE, DE = BE*

*Prove:* *AB = CD*

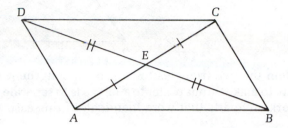

When attempting to write a proof, we do not just start writing down
steps, hoping that they will lead us to the conclusion! The secret lies in
the analysis—which starts with the conclusion and works backward up
the string of conditional statements until we reach a statement that was
given. If we wish to prove "if *A*, then *F*", we decide *F* will be true, if
*E* is; and *E* will be true, if *D* is; and *D* will be true, if *C* is; and *C* will
be true, if *B* is; and *B* will be true, if *A* is; and *A* was given.

$$A \rightarrow B$$
$$B \rightarrow C$$
$$C \rightarrow D$$
$$D \rightarrow E$$
$$E \rightarrow F$$
$$\therefore A \rightarrow F$$

**Starting with the conclusion
we work backward to the given.**

So in the exercise about the quadrilateral we decide *AB = CD* if they are
corresponding parts of congruent triangles. But which triangles? It
could be $\triangle ABD \cong \triangle CBD$, $\triangle ABC \cong \triangle ACD$, or $\triangle ABE \cong \triangle CDE$. Marking the given information on the figure with little lines seems to indicate
the most information about $\triangle ABE$ and $\triangle CDE$; but in order to prove
them congruent, we still need the included angle—but inspection
shows that the included angles are vertical angles! So, the line segments
are equal if they are corresponding parts of congruent triangles, the
triangles are congruent if we can find SAS, the sides were given equal,
and the angles are vertical angles. The body of the proof can then be
written:

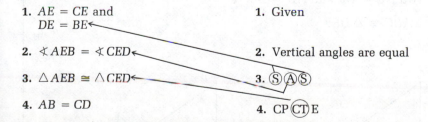

**1.** *AE = CE* and
    *DE = BE*

**2.** $\measuredangle AEB = \measuredangle CED$

**3.** $\triangle AEB \cong \triangle CED$

**4.** *AB = CD*

**1.** Given

**2.** Vertical angles are equal

**3.** Ⓢ Ⓐ Ⓢ

**4.** CP Ⓒ Ⓣ E

Notice that each reason is a conditional statement which has a
hypothesis—which must be established in order for the reason to be
valid (see circles and arrows above).

Another example:

Given: $\sphericalangle BAC = \sphericalangle ABC$
$\qquad AD = BE$

Prove: $\triangle ABD \cong \triangle BAE$

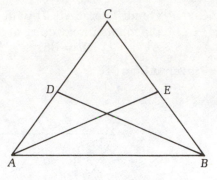

In this problem the two triangles to be proven congruent are overlapping. One way to make things easier to analyze is to separate them, being careful to mark any parts that were identical. In this case $AB = AB$.

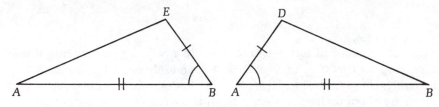

The proof should then be easy to see:

| | |
|---|---|
| **1.** $\sphericalangle BAC = \sphericalangle ABC$ | **1.** Given |
| **2.** $AD = BE$ | **2.** (1) |
| **3.** $AB = AB$ | **3.** Reflexive property (or "Identity") |
| **4.** $\triangle ABD \cong \triangle BAE$ | **4.** SAS |

The following theorem is usually presented as a corollary to the SAS theorem as it is easily derived from it:

---

*If two right triangles have their legs respectively equal, then the triangles are congruent.*

---

Given: Right triangles $ABC$ and $DEF$
$\qquad$ with $AC = DF$ and $AB = DE$

Prove: $\triangle ABC \cong \triangle DEF$

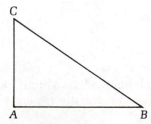
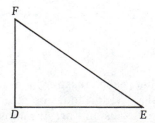

| | |
|---|---|
| **1.** $\triangle ABC$ and $\triangle DEF$ are right triangles | **1.** Given |
| **2.** $\angle BAC$ and $\angle EDF$ are right angles | **2.** Definition of right triangle |
| **3.** $\angle BAC = \angle EDF$ | **3.** All right angles are equal |
| **4.** $AC = DF$ | **4.** (1) |
| **5.** $AB = DE$ | **5.** (1) |
| **6.** $\triangle ABC \cong \triangle DEF$ | **6.** SAS |

As another illustration of the use of congruency, we will prove that the bisector of the vertex angle of an isosceles triangle is the perpendicular bisector of the base. However, it is easier to do so if we first have the following corollary to the definition of perpendicular:

> *If two lines intersect to form equal adjacent angles, then they are perpendicular.*

Given:  $\angle 1 = \angle 2$

Prove:  $AB \perp CD$

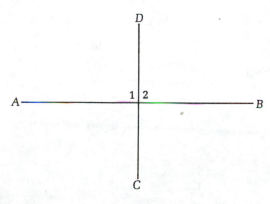

| | |
|---|---|
| **1.** $\angle 1 = \angle 2$ | **1.** Given |
| **2.** $\angle 1 + \angle 2$ form a straight angle | **2.** $AB$ is a straight line |
| **3.** $\angle 1 + \angle 2 = 180°$ | **3.** Definition of a straight angle |
| **4.** $\angle 1 + \angle 1 = 180°$ | **4.** Substitution |
| **5.** $2 \angle 1 = 180°$ | **5.** (4) |
| **6.** $\angle 1 = 90°$ | **6.** Division |
| **7.** $AB \perp CD$ | **7.** Definition of $\perp$ |

And now the theorem:

> *The bisector of the vertex angle of an isosceles triangle is the*
> *perpendicular bisector of the base.*

Given: Isosceles △ABC with
   CD bisecting ∡ACB

Prove: CD is the perpendicular
   bisector of AB

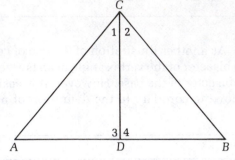

| 1. △ABC is isosceles | 1. Given |
|---|---|
| 2. AC = BC | 2. Def. of isosceles |
| 3. CD bisects ∡ACB | 3. (1) |
| 4. ∡1 = ∡2 | 4. Def. of bisect |
| 5. CD = CD | 5. Identity |
| 6. △ACD ≅ △BCD | 6. SAS |
| 7. AD = BD | 7. CPCTE |
| 8. D bisects AB | 8. (4) |
| 9. ∡3 = ∡4 | 9. (7) |
| 10. CD ⊥ AB | 10. If two lines intersect to form = adjacent ∡s they are ⊥ |
| 11. CD is ⊥ bisector of AB | 11. Def. of ⊥ bisector. |

Let us now proceed to the second of those three methods of proving
that triangles are congruent (SAS, ASA, and SSS).

> *If two angles and the included side of one triangle are equal
> respectively to two angles and the included side of another
> triangle, then the triangles are congruent.* (ASA)

Given: $\angle A = \angle D$,
$\quad\quad\quad AB = DE$,
$\quad\quad\quad \angle B = \angle E$

Prove: $\triangle ABC \cong \triangle DEF$

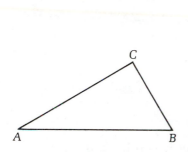 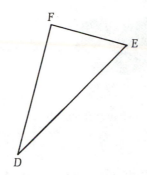

1. Rotate triangle *DEF* so that *E* coincides with *B* and *DE* falls on *AB*

   1. Existence of a rotation

2. $AB = DE$

   2. Given

3. *D* will coincide with *A*

   3. Distance Postulate

4. If *C* and *F* are on opposite sides of *AB*, reflect $\triangle D$ about *AB*

   4. Existence of a reflection

5. $\angle E = \angle B$

   5. (2)

6. *EF* will coincide with *BC*

   6. Angle Measure Postulate

7. $\angle A = \angle D$

   7. (2)

8. *DF* will coincide with *AC*

   8. (6)

9. $\triangle ABC \cong \triangle DEF$

   9. Def. of congruent; that is, they coincide throughout and all the corresponding parts are equal

*"So we see that these triangles are not merely similar, they are actually congruent! Isn't that wonderful? Doesn't it make you want to sing and shout?"*

© Punch (Rothco).

The following theorem is an easy corollary to ASA:

> **If two right triangles have a leg and adjacent angle respectively equal, then they are congruent.**

Given:  Right triangles $ABC$ and $DEF$
with $AC = DF$ and $\angle C = \angle F$

Prove:  $\triangle ABC \cong \triangle DEF$

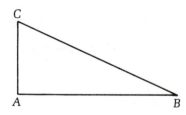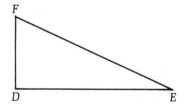

| | |
|---|---|
| 1. $\triangle BAC$ and $\triangle EDF$ are rt. $\triangle$s | 1. Given |
| 2. $\angle BAC$ and $\angle EDF$ are rt. $\angle$s | 2. Def. of rt. $\triangle$ |
| 3. $\angle ABC = \angle DEF$ | 3. All rt. $\angle$s are = |
| 4. $\angle C = \angle F$ | 4. Given |
| 5. $AC = DF$ | 5. (4) |
| 6. $\triangle ABC \cong \triangle DEF$ | 6. ASA |

Now some examples which make use of congruency by ASA:

**Example 1**

Given: $AD = BF$
$\angle 2 = \angle 6$
$\angle 3 = \angle 4$

Prove: $\triangle ABC \cong \triangle DEF$

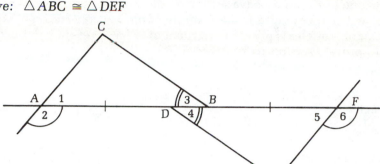

(Analysis: By marking the given information on the figure, we see that one pair of angles is equal. If we are to have congruency by ASA, we need another pair of angles and the included side. The required angles are supplements to angles that are given as equal, and the included sides can be formed by adding $DB$ to each of the line segments that were given equal.)

1. $AD = BF$ | 1. Given
2. $DB = DB$ | 2. Reflexive
3. $AB = DF$ | 3. Addition
4. $\angle 2 = \angle 6$ | 4. (1)
5. $\angle 1$ and $\angle 2$ are supplementary and $\angle 5$ and $\angle 6$ are supplementary | 5. Def. of supplementary
6. $\angle 1 = \angle 5$ | 6. Supplements of equal angles
7. $\angle 3 = \angle 4$ | 7. (1)
8. $\triangle ABC \cong \triangle DEF$ | 8. ASA

**Example 2**

Given: $\angle ADC$, $\angle ABC$, $\angle CED$, and $\angle AFB$
are right angles, $\angle 1 = \angle 2$,
$AF = CE$

Prove: $AD = BC$

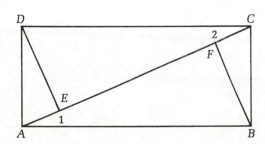

(Analysis: $AD = BC$ if they are corresponding parts of congruent triangles. There are two pairs of possible congruent triangles—$\triangle ABC \cong \triangle ADC$ or $\triangle BCF \cong \triangle ADE$. But we do not have enough given information to prove either of these pairs congruent. We need to derive some more information. Since, in the first named pair, we have two angles equal to two angles, we would like to have $AB = CD$. They are parts of $\triangle ABF$ and $CDE$, and we have sufficient information given to prove them congruent. This exercise is an example of *double congruency*—where one pair of triangles is proven congruent in order to get enough information to prove a second pair congruent.)

| | |
|---|---|
| **1.** $\measuredangle CED$ and $\measuredangle AFB$ are rt. angles | **1.** Given |
| **2.** $\measuredangle CED = \measuredangle AFB$ | **2.** All rt. angles are equal |
| **3.** $AF = CE$ | **3.** (1) |
| **4.** $\measuredangle 1 = \measuredangle 2$ | **4.** (1) |
| **5.** $\triangle CED \cong \triangle AFB$ | **5.** ASA |
| **6.** $AB = CD$ | **6.** CPCTE |
| **7.** $\measuredangle ADC$ and $\measuredangle ABC$ are rt. angles | **7.** (1) |
| **8.** $\measuredangle ADC = \measuredangle CBA$ | **8.** (2) |
| **9.** $\triangle ADC \cong \triangle ABC$ | **9.** ASA |
| **10.** $AD = BC$ | **10.** (6) |

NOTE: There are many other ways that this could have been proven. None is any more correct than another. The shortest proof is not necessarily the best one, either, as a longer proof may be more easily followed by the reader. Can you see other ways of proving the example above?

**EXERCISE SET 9**

1. Given: $BD$ bisects $\measuredangle B$
   $\quad\quad\quad AB = BC$

   Prove: $\triangle ABD \cong \triangle CDB$

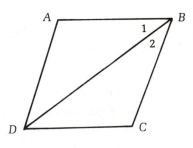

2. Given: $AC$ and $BD$ bisect each other

   Prove: $\triangle ABO \cong \triangle CDO$

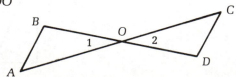

3. Given: $OB = OC$
   $\angle 1 = \angle 2$

   Prove: $\angle B = \angle C$

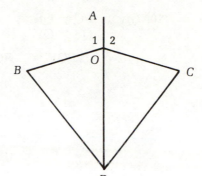

4. Given: $AE = AD$
   $BE = CD$
   Prove: $EC = BD$
   Hint: Use overlapping triangles

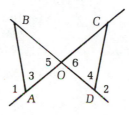

5. Given: $\angle 1 = \angle 2$
   $AO = OD$

   Prove: $AB = CD$

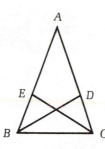

6. Given: $AD$ bisects $\angle BAC$
   $AD \perp BC$

   Prove: $\triangle ABC$ is isosceles

Figure for problems 6 through 8.

7. Given: $AD$ is $\perp$ bisector of $BC$

   Prove: $\triangle ABC$ is isosceles

8. Given: $\triangle ABC$ is isosceles
   $\angle 1 = \angle 2$

   Prove: $D$ bisects $BC$

9. Given: $DF$ is $\perp$ bis. of $AB$
   $EF$ is $\perp$ bis. of $BC$

   Prove: $AF = CF$

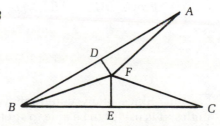

10. Given: PQ = QR
      QS = QV

   Prove: △ PQV ≅ △ RQS

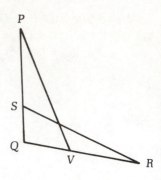

11. Given: ∢1 = ∢2
      AC = BD

   Prove: △ ACD ≅ △ BDC

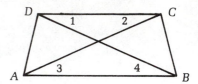

12. Given: MN and PQ are straight lines
      intersecting at O
      OM = ON
      ∢ M = ∢ N

   Prove: △ MOP ≅ △ NOQ

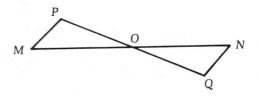

13. Given: ∢ ABC = ∢ ACB
      BD bisects angle ABC
      CE bisects angle ACB

   Prove: BD = CE

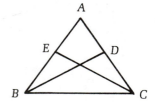

*14. Given: ∢1 = ∢2
       ∢3 = ∢4

    Prove: AD = AE

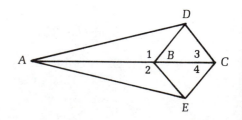

15. Since many congruency problems involve overlapping triangles, it is useful to be able to spot them. How many triangles are there

in this figure? If you organize carefully, you can use the symmetry to simplify your counting.

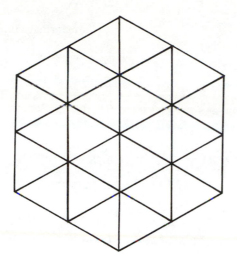

Before we prove the third of our major methods of congruency, it will be necessary to have the following theorem about isosceles triangles:

**10. Isosceles triangles and SSS-congruent triangles**

> *If two sides of a triangle are equal, the angles opposite those sides are equal.*

This is sometimes also stated as, "The base angles of an isosceles triangle are equal."

Given: △ABC with AB = BC

Prove: ∡A = ∡C

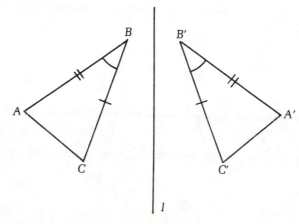

*1*

1. Reflect $\triangle ABC$ about any line $l$
2. $AB = BC$
3. $AB = A'B'$
4. $BC = A'B'$
5. $BC = B'C'$
6. $AB = B'C'$
7. $\measuredangle B = \measuredangle B'$
8. $\triangle ABC \cong \triangle A'B'C'$
9. $\measuredangle A = \measuredangle C'$
10. $\measuredangle C = \measuredangle C'$
11. $\measuredangle A = \measuredangle C$

1. Existence of reflections
2. Given
3. Transformation invariance
4. Substitution
5. (3)
6. (4)
7. (3)
8. SAS
9. CPCTE (see note below)
10. (3)
11. (4)

NOTE the importance in this proof of being careful about which parts are the corresponding parts.

The converse of this theorem is also true, and we will prove it at this time:

> *If two angles of a triangle are equal then the sides opposite those angles are equal.*

Given: $\triangle ABC$ with $\measuredangle A = \measuredangle C$

Prove: $AB = BC$

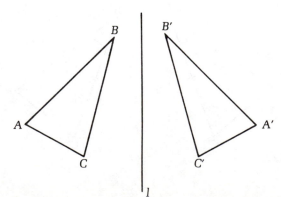

| | |
|---|---|
| 1. Reflect $\triangle ABC$ about any line $l$ | 1. Existence of reflections |
| 2. $\measuredangle A = \measuredangle C$ | 2. Given |
| 3. $\measuredangle C' = \measuredangle C$ | 3. Transformation invariance |
| 4. $\measuredangle A = \measuredangle C'$ | 4. Substitution |
| 5. $\measuredangle A = \measuredangle A'$ | 5. (3) |
| 6. $\measuredangle C - \measuredangle A'$ | 6. (4) |
| 7. $AC = A'C'$ | 7. (3) |
| 8. $\triangle ABC \cong \triangle A'B'C'$ | 8. ASA |
| 9. $AB = B'C'$ | 9. CPCTE |
| 10. $BC = B'C'$ | 10. (3) |
| 11. $AB = BC$ | 11. (4) |

These two theorems about isosceles triangles have corollaries which extend the concept involved to equilateral triangles:

---

### *An equilateral triangle is also equiangular.*

Given: Equilateral $\triangle ABC$

Prove: $\triangle ABC$ is equiangular

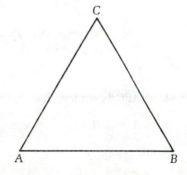

| | |
|---|---|
| 1. $\triangle ABC$ is equilateral | 1. Given |
| 2. $AB = BC$ | 2. Def. of equilateral |
| 3. $\measuredangle A = \measuredangle C$ | 3. $\measuredangle$s opposite = sides are = |
| 4. $BC = AC$ | 4. (2) |
| 5. $\measuredangle A = \measuredangle B$ | 5. (3) |
| 6. $\measuredangle B = \measuredangle C$ | 6. Substitution |
| 7. $\triangle ABC$ is equiangular | 7. Def. of equiangular |

And also, the theorem from which this corollary comes has a converse whose corollary is the converse of this corollary. (Oh, yeah?!) Maybe we had better draw a picture:

| 2 sides = → opposite ∡s = | – converse – | 2 ∡s = → opposite sides = |
|:---:|:---:|:---:|
| &#124; | | &#124; |
| corollary | | corollary |
| &#124; | | &#124; |
| equilateral → equiangular | – converse – | equiangular → equilateral |

At any rate:

---

### An equiangular triangle is also equilateral.

*Given:* △ABC is equiangular

*Prove:* △ABC is equilateral

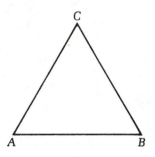

| 1. △ABC is equiangular | 1. Given |
|---|---|
| 2. ∡A = ∡B | 2. Def. of equiangular |
| 3. AC = BC | 3. Sides opposite = ∡ s are = |
| 4. ∡B = ∡C | 4. (2) |
| 5. AB = AC | 5. (3) |
| 6. AB = BC | 6. Substitution |
| 7. △ABC is equilateral | 7. Def. of equilateral |

---

Now to prove our third major congruency theorem:

> *If three sides of one triangle are equal respectively to three
> sides of another triangle, then the triangles are congruent.*
> **(SSS)**

Given: $AB = DE$
$BC = EF$
$AC = DF$

Prove: $\triangle ABC \cong \triangle DEF$

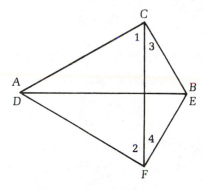 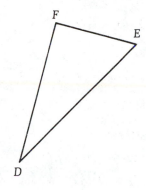

| | |
|---|---|
| **1.** Rotate & reflect $\triangle DEF$ so that $E$ coincides with $B$ and $DE$ falls on $AB$ | **1.** Existence of reflections and rotations |
| **2.** $AB = DE$ | **2.** Given |
| **3.** $D$ will coincide with $A$ | **3.** Distance Postulate ($AB$ and $DE$ represent the same real number) |
| **4.** If $C$ and $F$ are on the same side of $AB$, reflect $\triangle DEF$ about $AB$ | **4.** Existence of a reflection |
| **5.** Draw $CF$ | **5.** Two points determine a line |
| **6.** $AC = DF$ | **6.** (2) |
| **7.** $\angle 1 = \angle 2$ | **7.** $\angle$s opposite = sides of a $\triangle$ are = |
| **8.** $BC = EF$ | **8.** (2) |
| **9.** $\angle 3 = \angle 4$ | **9.** (7) |
| **10.** $\angle ACB = \angle DFE$ | **10.** Addition |
| **11.** $\triangle ABC \cong \triangle DEF$ | **11.** SAS |

Now let us do a couple of very simple examples to illustrate the use of the SSS theorem. The second example makes use of overlapping triangles.

**Example 1:**

Given: $AB = CD$
$\qquad BC = AD$

Prove: $\angle 1 = \angle 2$

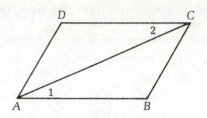

1. $AB = CD$     1. Given
2. $BC = AD$     2. (1)
3. $AC = AC$     3. Reflexive
4. $\triangle ABC \cong \triangle CDA$     4. SSS
5. $\angle 1 = \angle 2$     5. CPCTE

**Example 2:**

Given: $AE = BD$
$\qquad AD = BE$

Prove: $AC = BC$

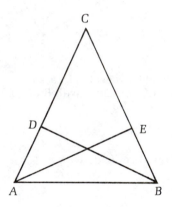

1. $AE = BD$     1. Given
2. $AD = BE$     2. (1)
3. $AB = AB$     3. Reflexive
4. $\triangle ABD \cong \triangle ABE$     4. SSS
5. $\angle CAB = \angle CBA$     5. CPCTE
6. $AC = BC$     6. Sides opposite = angles are =

1. Given: $AB = AC$

   Prove: $\angle 3 = \angle 4$

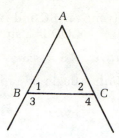

2. Given: $ABC$ is an isosceles triangle

   $E$ and $D$ are midpoints of the equal sides

   Prove: $BD = CE$

   Hint:　Use the lower overlapping triangles

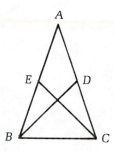

3. Given: $\angle CAB = \angle CBA$

   $\angle 1 = \angle 2$

   Prove: $AP = BP$

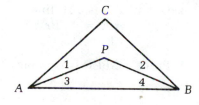

4. Given: $\angle DAB = \angle DBA$

   $\angle 1 = \angle 2$

   Prove: $\angle 5 = \angle 6$

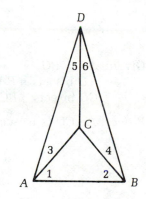

5. Given: $AB = CD$

   $AD = BC$

   Prove: $\angle A = \angle C$

   Hint:　Draw $BD$

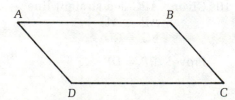

6. Given: △*ABC* and △*ABO* are isosceles with common base *AB*
   *COD* is a straight line

   Prove: ∢5 = ∢6

   Hint: Begin by proving the upper triangles congruent.

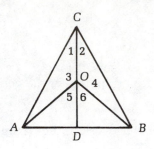

7. Given: *AB* = *BC*
   *CD* = *AD*

   Prove: ∢*BAD* = ∢*BCD*

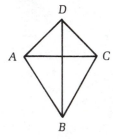

8. Given: *AD* is a st. line
   ∢*A* = ∢*CBF*
   *CE* = *AE*

   Prove: △*BCF* is isosceles

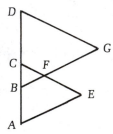

9. Given: *AB* = *BC*
   *AD* bisects ∢*BAC*
   *CD* bisects ∢*BCA*

   Prove: △*ACD* is isosceles

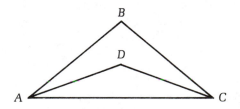

10. Given: *APC* is a straight line
    *AB* = *AD*
    *BC* = *CD*

    Prove: *BP* = *DP*

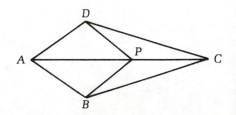

11. Prove the median to the base of an isosceles triangle bisects the vertex angle. (The median is the line from a vertex to the mid-point of the opposite side.)

12. Prove the median to the base of an isosceles triangle is perpendicular to the base.

We now have a sufficient supply of theorems to look again at the basic constructions discussed earlier and *prove* that the manipulations we go through will always accomplish the desired result.

The use of the straight-edge to draw a line is justified by the postulate that two points determine a line. The use of the compass is justified by the definition of a circle—the set of all points that are a fixed distance from a given point called the center. When we mark an arc on a line using a given setting of the compass, we are using radii of the same circle that, by the definition of a circle, would be equal.

In what follows, we will not repeat the discussion of the technique, but will only give the resulting figure and the body of the proof.

## 11. Proofs for the constructions, auxiliary lines

*I hear and I forget.*
*I see and I remember.*
*I do and I understand.*
            *—Chinese Proverb*

To **copy a line segment:**

$AB = CD$                                    Radii of the same circle

To **copy an angle:**

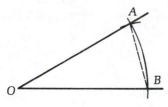   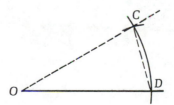

**1.** $AO = BO = CO = DO$

**2.** $AB = CD$

**3.** $\triangle AOB \cong \triangle COD$

**4.** $\angle AOB = \angle COD$

**1.** Radii of the same circle are =

**2.** (1)

**3.** SSS

**4.** CPCTE

To **bisect a line segment:**

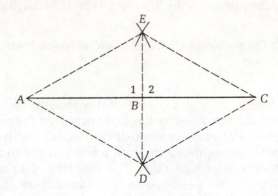

| 1. $AD = AE = CD = CE$ | 1. Radii of the same circle are = |
|---|---|
| 2. $DE = DE$ | 2. Reflexive |
| 3. $\triangle ADE \cong \triangle CDE$ | 3. SSS |
| 4. $\sphericalangle AEB = \sphericalangle CEB$ | 4. CPCTE |
| 5. $BE = BE$ | 5. Reflexive |
| 6. $\triangle ABE \cong \triangle CBE$ | 6. SAS |
| 7. $AB = BC$ | 7. CPCTE |
| 8. $B$ bisects $AC$ | 8. Def. of bisector |

We can continue the proof to show that $DE$ is not only the bisector of $AC$, but is perpendicular to it as well:

| 9. $\sphericalangle 1 = \sphericalangle 2$ | 9. CPCTE |
|---|---|
| 10. $DE \perp AC$ | 10. If 2 lines intersect to form = adjacent angles, they are $\perp$ |

To **bisect an angle:**

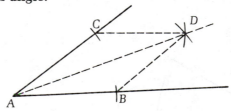

| 1. $AC = AB$ | 1. Radii of the same circle are = |
|---|---|
| 2. $BD = CD$ | 2. (1) |
| 3. $AD = AD$ | 3. Reflexive |
| 4. $\triangle ABD \cong \triangle ACD$ | 4. SSS |
| 5. $\sphericalangle CAD = \sphericalangle BAD$ | 5. CPCTE |
| 6. $AD$ bisects $\sphericalangle CAB$ | 6. Def. of bisector |

To construct a **perpendicular to a line at a point on the line:**

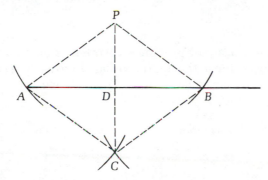

| | |
|---|---|
| **1.** $PA = PB$ | **1.** Radii of the same circle are = |
| **2.** $AC = BC$ | **2.** (1) |
| **3.** $CP = CP$ | **3.** Reflexive |
| **4.** $\triangle PCA \cong \triangle PCB$ | **4.** SSS |
| **5.** $\angle 1 = \angle 2$ | **5.** CPCTE |
| **6.** $CP \perp AB$ | **6.** If 2 lines intersect to form = adjacent angles, they are $\perp$ |

Notice that the construction of a perpendicular to a line at a point on a line is a special case of the construction of an angle bisector. When we bisect the straight angle, we get a perpendicular.

To construct a **perpendicular to a line from a point not on the line:**

| | |
|---|---|
| **1.** $PA = PB$ | **1.** Radii of the same circle are = |
| **2.** $AC = BC$ | **2.** (1) |
| **3.** $PC = PC$ | **3.** Reflexive |
| **4.** $\triangle PCA = \triangle PCB$ | **4.** SSS |
| **5.** $\angle APD = \angle BPD$ | **5.** CPCTE |
| **6.** $PD = PD$ | **6.** Reflexive |
| **7.** $\triangle PDA \cong \triangle PDB$ | **7.** SAS |
| **8.** $\angle PDA = \angle PDB$ | **8.** CPCTE |
| **9.** $PD \perp AB$ | **9.** If 2 lines intersect to form = adjacent angles they are $\perp$ |

Before the next construction can be proven, "parallel" needs to be defined and the theorem about corresponding angles must be proved. Nevertheless, we will present the "proof" here for the sake of completeness.

To construct a **parallel to a given line through a given point:**
(Refer to the construction on page 50)

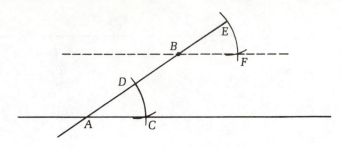

| | |
|---|---|
| **1.** $AC = AD = BE = BF$ | **1.** Radii of the same circle are = |
| **2.** $CD = EF$ | **2.** (1) |
| **3.** $\triangle ACD \cong \triangle BFE$ | **3.** SSS |
| **4.** $\angle CAD = \angle FBE$ | **4.** CPCTE |
| **5.** $BF \parallel AC$ | **5.** Corresponding angles are equal (a method for proving lines parallel, which is proven in the section on parallel lines). |

If, in the process of proving a theorem or an exercise, it would be helpful to add a line to the figure, we can do so as long as it can be added by using one of our constructions. For instance:

Draw $AE$ and $BD$ bisectors of     Construction
$\angle A$ and $\angle B$

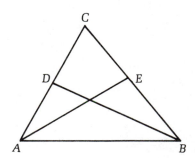

Draw *DE* and *BF* ⊥ to *AC*          Construction          **99**

**11–PROOFS FOR
THE
CONSTRUCTIONS,
AUXILIARY
LINES**

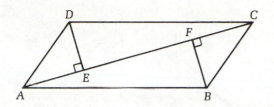

Be careful, however, *not* to do something like:

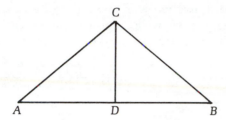

Draw *CD* ⊥ bis. of *AB*                    Construction

This is requiring more than can be done, except in special cases. We do not have a construction for a perpendicular bisector of a given line, passing through a given point not on the line. The perpendicular bisector of *AB* might not pass through *C*:

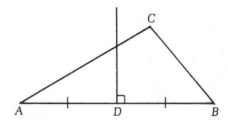

Or the line from *C* to the midpoint of *AB* might not be perpendicular to *AB*:

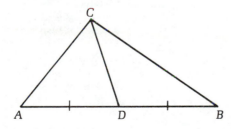

Or the perpendicular from $C$ to $AB$ might not bisect $AB$:

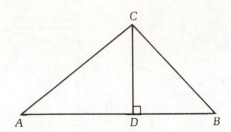

In the proof that follows we will make use of an **auxiliary** line. This theorem and the next one have been proven previously, but this time, we will prove them without the use of the transformation concepts.

> *If two sides of a triangle are equal, then the angles opposite those sides are equal.*

Given: $\triangle ABC$ with $AB = BC$

Prove: $\angle A = \angle C$

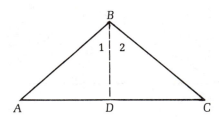

| | |
|---|---|
| **1.** Draw $BD$ bisecting $\angle B$ | **1.** Construction |
| **2.** $AB = BC$ | **2.** Given |
| **3.** $\angle 1 = \angle 2$ | **3.** Def. of bisector |
| **4.** $BD = BD$ | **4.** Reflexive |
| **5.** $\triangle ABD \cong \triangle CBD$ | **5.** SAS |
| **6.** $\angle A = \angle C$ | **6.** CPCTE |

> *If two angles of a triangle are equal, then the sides opposite those angles are equal.*

This proof without transformations will be much more difficult. It not only uses auxiliary lines, but uses double congruency. No beginning geometry student would be expected to derive this proof on his own, but that student should be able to follow the completed proof.

Given: $\triangle ABC$ with $\angle A = \angle C$

Prove: $AB = BC$

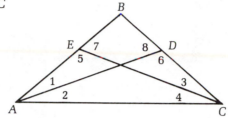

| | |
|---|---|
| 1. Draw $AD$ bisecting $\angle A$ and $CE$ bis. $\angle C$ | 1. Construction |
| 2. $\angle A = \angle C$ | 2. Given |
| 3. $\angle 2 = \angle 4$ | 3. Halves of equals are equal |
| 4. $AC = AC$ | 4. Reflexive |
| 5. $\triangle ACE \cong \triangle ACD$ | 5. ASA |
| 6. $\angle 5 = \angle 6$ | 6. CPCTE |
| 7. $\angle 5$ and $\angle 7$ are supplementary | 7. Given $ABC$ is a triangle, that is, $AB$ is a st. line |
| 8. $\angle 6$ and $\angle 8$ are supplementary | 8. Given $ABC$ is a triangle, that is, $BC$ is a st. line |
| 9. $\angle 7 = \angle 8$ | 9. Supp. of equals are equal |
| 10. $AD = CE$ | 10. CPCTE |
| 11. $\angle 3 = \angle 1$ | 11. Halves of equals are equal |
| 12. $\triangle BEC \cong \triangle BDA$ | 12. ASA |
| 13. $AB = BC$ | 13. CPCTE |

During the Middle Ages the proof of *If two sides of a triangle are equal the opposite angles are equal* became known as "The Bridge of Fools". When the proof is done as Euclid did it, the auxiliary lines make the figure look like a bridge with its trussing. Also, many students were not able to do this proof and thus were not able to "cross the bridge" to the rest of the geometry course. The proof goes like this:

Given: $\triangle ABC$ with $AC = BC$

Prove: $\angle CAB = \angle CBA$

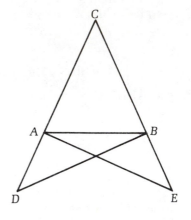

1. Extend $CA$ to some point $D$    1. Construction

2. Extend $CB$ to $E$ so that $AD = BE$    2. Construction (copy a line)

3. Draw $AE$ and $BD$    3. Construction

4. $AC = BC$    4. Given

5. $CD = CE$    5. Addition

6. $\angle C = \angle C$    6. Reflexive

7. $\triangle CDB \cong \triangle CEA$    7. SAS

8. $BD = AE$    8. CPCTE

9. $\angle D = \angle E$    9. CPCTE

10. $\triangle ABD \cong \triangle ABE$    10. SAS

11. $\angle BAD = \angle ABE$    11. CPCTE

12. $\angle BAD$ supp. to $\angle CAB$    12. $CD$ is st. line

13. $\angle ABE$ supp. to $\angle CBA$    13. $CE$ is st. line

14. $\angle CAB = \angle CBA$    14. Supp. of = angles are =

NOTE: we have now proven this theorem three different ways!

1. Given: $AD = BC$, $AB = CD$

   Prove: $\angle A = \angle C$

   (Hint: Draw BD)

2. Given: $AC = BC$, $\angle 1 = \angle 2$

   Prove: $\angle 3 = \angle 4$

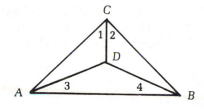

3. Given: D, E, F are midpoints of $AB$, $BC$, and $AC$ respectively, $AC = BC$

   Prove: $\angle A = \angle B$

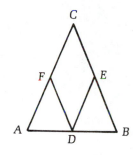

4. Given: $AO = OB$, $DO = OC$

   Prove: $EO = OF$

   (Hint: First prove $\triangle AOD \cong \triangle BOC$.)

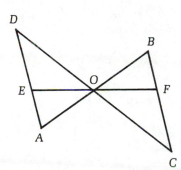

5. Given: $AB = CD$, $AD = BC$, $PQ$ bisects $DB$ at $O$

   Prove: $PO = QO$

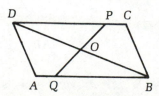

6. Given: $\angle 1 = \angle 2$, $\angle 3 = \angle 4$, $AE = CF$

   Prove: $AB = CD$

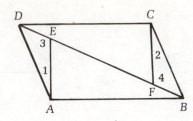

7. Prove that when constructing the reflection of point $A$ about line $L$, $A'$ will always be at the intersection of the construction marks when the compass setting is not changed.

   Given: $AC = AB = A'B = A'C$

   Prove: $BC$ is $\perp$ bisector of $AA'$

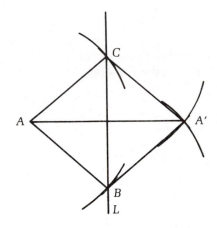

8. Given: Same as exercise 3

   Prove: $\angle FDA = \angle EDB$

9. Given: Same figure as exercise 4
   $DO = OC$
   $EO = OF$

   Prove: $\angle A = \angle B$

10. Given: Same figure as exercise 5
    $AB = CD$
    $OP = OQ$
    $PQ$ bisects $BD$ at $O$

    Prove: $\angle A = \angle C$

11. Given: Same figure as exercise 6
    $AD = BC$
    $AB = CD$
    $DE = BF$

    Prove: $\angle 1 = \angle 2$

**105**

**11–PROOFS FOR
THE
CONSTRUCTIONS,
AUXILIARY
LINES**

12. Given:   Same as exercise 6
    Prove:   $DF = BE$

13. Given:   Same figure as exercise 6
    $AD = BC$
    $AB = CD$
    $DE = BF$

    Prove:   $\angle BAE = \angle DCF$

14. Prove:   Any point on the perpendicular bisector of a line segment is equidistant from the ends of the segment.

15. Given:   M is the midpoint of $AB$
    $AD = BC$
    $\angle A = \angle B$

    Prove:   M is equidistant from C and D

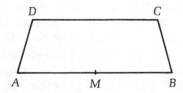

16. Given:   $AB = AD$
    $BC = CD$

    Prove:   $\angle B = \angle D$

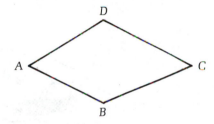

17. Given:   $PQ = RS$
    $QR = PS$

    Prove:   $\angle Q = \angle S$

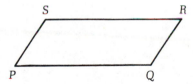

18. Given:   $AD = BC$
    $AC = BD$

    Prove:   $\angle C = \angle D$

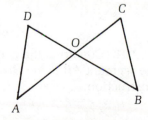

19. The story is told that Napoleon, on one of his marches, came upon a river and wanted to know the distance across it. Since his engineers were in the rear, he asked a junior officer the distance across. The officer stood erect, and pulled the visor of his cap down until he could just see the far bank of the river. He then turned and sighted along

the visor edge to a landmark on his side of the river. He then paced off the distance to this landmark and told Napoleon that this was the distance across the river. (The story is probably not true, as Napoleon was an accomplished mathematician.) Prove that the paced-off distance is the distance across the river.

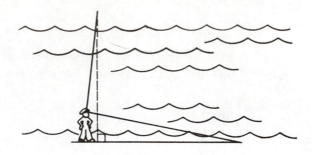

20. The following is another method of measuring an inaccessible distance. Prove that it works. To find the distance, *AB*, across the lake, go at right angles to *AB* any convenient distance to *D*. Place a marker at *D* and continue on the same distance to *C*. Turn at right angles again and go away from the lake until you reach a point *E* where your marker at *D* lines up with the point *B* on the far side of the lake. *CE* will then be the same distance as the distance across the lake.

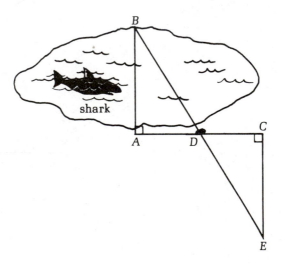

*21. Construct the perpendicular bisector of line segment *AB*. The line segment *AB* is too close to the bottom of the paper to use the usual construction.

*22. Construct the perpendicular to line $AB$ at $P$. Point $P$ is too close to the edge of the paper to use the usual construction.

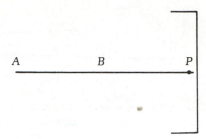

## 12. Parallel lines

We will define **parallel lines** as lines in the same plane that do not intersect, no matter how far they are extended. Notice that we specify the lines be in the same plane. Lines in different planes that do not intersect are called **skew lines**—such as the edges $AB$ and $CD$ of this cube:

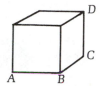

Parallel lines do not intersect? *Delaware and Hudson Railroad.*

This definition agrees well with our intuitive ideas about parallels, and will be sufficient for our work here, but is not the best mathematical definition, nor is it very realistic. Are these two lines parallel?

They do not meet. You would probably claim that they would meet if they were extended. Suppose two lines are drawn all the way across a blackboard and do not meet. Are they parallel?

You would probably say no—they will meet if extended further. Given another pair of lines drawn all the way across the blackboard, would you say these are parallel?

If you say yes, how would you go about proving it? If you say no, how would you go about disproving it? If you extend a pair of lines 100 m and they do not intersect, does that prove them parallel? How about 1000 km, or from here to the nearest star, or to the farthest star? So you see, our definition cannot be used to test lines and determine whether or not they are parallel. We will have to accept the definition as a concept, and derive some theorems to use when we wish to determine whether or not two lines are parallel.

Since we cannot use the definition for developing a direct proof, we will have to resort to other methods—called indirect proofs. Basically there are three types:

*Reductio ad absurdum*—which translates from the Latin as "reduce to an absurdity." This method involves situations where there are exactly two possibilities. Two lines are either perpendicular, or they are not perpendicular. A number is either rational or irrational. Two lines are parallel or they are not parallel. If, by some manner of reasoning, we can reduce one of the two possibilities to an absurdity, then the other possibility *must* be true. We will use this method in some of the proofs which follow.

As an example, let us prove a previous theorem with an indirect proof:

Given: $\angle A = \angle B$

Prove: $AC = BC$

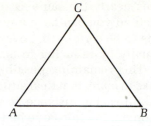

**1.** Construct the $\perp$ bisector of $AB$     **1.** Construction

The $\perp$ bisector either passes through C, or it does not pass through C. Assume it does *not* pass through C:

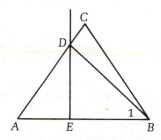

**2.** Draw $BD$               **2.** Construction

**3.** $AE = BE$           **3.** Def. of bisector

**4.** $\angle AED = \angle BED$      **4.** Def. of $\perp$

**5.** $DE = DE$          **5.** Reflexive

**6.** $\triangle AED \cong \triangle BED$     **6.** SAS

**7.** $\angle A = \angle EBD$       **7.** CPCTE

**8.** $\angle A = \angle B$         **8.** Given

**9.** $\angle EBD = \angle B$      **9.** Substitution

But this is a contradiction.     The whole is greater than any of its parts.

$\therefore$ the $\perp$ bisector of $AB$ passes through C.     The only other possibility has been eliminated. Reductio ad adsurdum.

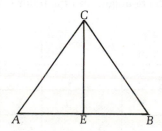

**10.** $\triangle AEC \cong \triangle BEC$     **10.** Same method as above

**11.** $AC = BC$           **11.** CPCTE

*Exclusion*—is nearly the same as *reductio ad adsurdum*, except that with this method you can have any number of possibilities. The technique here is to be *certain* that every possibility is listed, and then to exclude (usually by reducing to an absurdity) all of the possibilities except one. This remaining possibility *must* be true. This method is difficult to use in many practical situations owing to the difficulty of listing all possibilities. For instance, in a law case, with three suspects, you would not convict the third one just because the first two had ironclad alibis. However, it would be valid reasoning to convict the third suspect if you could be *certain* that the crime was committed by one of the three suspects. That is, that you had listed *all* the possibilities.

Sometimes it is useful, when using the indirect method of exclusion, to set up a table of all possibilities and eliminate those that are impossible.

As an example:

**In a certain bank the positions of cashier, manager, and teller are held by Brown, Jones, and Smith—though not necessarily in that order.**

**1. The teller, who is an only child, earns the least.**
**2. Smith, who is married to Brown's sister, earns more than the manager.**

**What position does each man hold?**

**By statements 1 and 2, Brown cannot be the teller.**

|         | B | J | S |
|---------|---|---|---|
| Cashier |   |   |   |
| Manager |   |   |   |
| Teller  | X |   |   |

**By statement 2, Smith is not the manager.**

|     | B | J | S |
|-----|---|---|---|
| C   |   |   |   |
| M   |   |   | X |
| T   | X |   |   |

**Smith earns more than the manager and the teller earns the least, so Smith is not the teller.**

|     | B | J | S |
|-----|---|---|---|
|     |   |   |   |
| M   |   |   | X |
| T   | X |   | X |

**Therefore, Smith is the cashier, and Jones and Brown are not the cashier.**

|   | B | J | S |
|---|---|---|---|
| C | X | X | O |
| M |   |   | X |
| T | X |   | X |

**Which identifies Brown as the manager, which in turn makes Jones the teller.**

|   | B | J | S |
|---|---|---|---|
| C | X | X | O |
| M | O | X | X |
| T | X | O | X |

*Coincidence*—is a method in which we make two figures coincide. When two figures coincide, each will have all the properties of the other. For instance, if we can show that the line through $C$, parallel to $AB$, coincides with the bisector, $EC$, of $\angle DCA$; then the angle bisector will be parallel to $AB$, and the line parallel to $AB$ through $C$ will bisect $\angle DCA$.

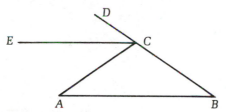

In the theorems which follow—methods of proving lines parallel—some of the reasons given are not stated exactly the same as the axioms, postulates, definitions, or previously proven theorems—but, in all cases, are statements which can be derived from them.

*B.C. by permission of Johnny Hart and Field Enterprises, Inc.*

> *Two lines in the same plane, perpendicular to the same line, are parallel.*

Given: $m \perp l$, $n \perp l$

Prove: $m \parallel n$

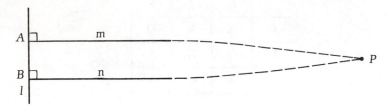

(Analysis: We will use *reductio ad absurdum*—assume they are not parallel, and show that this leads to a contradiction; therefore, they must be parallel.)

| | |
|---|---|
| **1.** Either $m \parallel n$ or $m \nparallel n$ | **1.** Two lines in the same plane either intersect or are parallel |
| **2.** Assume $m \nparallel n$ $m$ and $n$ will intersect at some point P | **2.** Non-parallels must intersect |
| **3.** $PA \perp l$ and $PB \perp l$ | **3.** Given |
| **4.** But this is impossible | **4.** From a given point a unique $\perp$ can be drawn to a given line (construction of $\perp$) |
| **5.** $\therefore m \parallel n$ | **5.** *Reductio ad absurdum* |

> *Two lines in the same plane, parallel to the same line, are parallel to each other.*

Given: $p \parallel r$, $q \parallel r$

Prove: $p \parallel q$

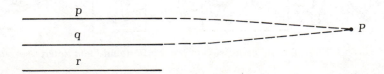

(Analysis: Again use *reductio ad absurdum*.)

1. Either $p \parallel q$ or $p \nparallel q$

1. Two lines in the same plane either intersect or are parallel

2. Assume $p \nparallel q$
   $p$ and $q$ will intersect at some point P

2. Non-parallels must intersect

3. $p \parallel r$ and $q \parallel r$

3. Given

4. But this is impossible

4. Through a given point only one parallel can be drawn to a given line

5. $\therefore p \parallel q$

5. *Reductio ad absurdum*

For the next few theorems and corollaries we need to use a **transversal**. Since trans means across, a transversal is a line which cuts across two or more other lines.

When a transversal cuts across two other lines, two sets of four angles are formed:

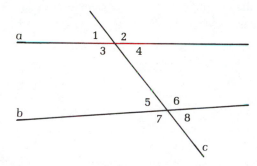

The angles between lines $a$ and $b$ are called **interior** angles ($\angle 3$, $\angle 4$, $\angle 5$, $\angle 6$). The angles outside of lines $a$ and $b$ are called **exterior** angles ($\angle 1$, $\angle 2$, $\angle 7$, $\angle 8$). Angles on opposite sides of the transversal $c$, with different vertices, are **alternate** angles. Putting some of these terms together, to compare the angles of each set, $\angle 3$ and $\angle 6$ are **alternate interior** angles, $\angle 4$ and $\angle 5$ are alternate interior angles, $\angle 1$ and $\angle 8$ are **alternate exterior** angles, $\angle 2$ and $\angle 7$ are also alternate exterior angles. If the upper set of four angles were placed on top of the lower set of four angles, the ones on top of each other would be called **corresponding** angles. There are four pairs of corresponding angles: $\angle 1$ and $\angle 5$, $\angle 2$ and $\angle 6$, $\angle 7$, $\angle 4$ and $\angle 8$.

*If two lines are cut by a transversal so that the alternate interior angles are equal, then the lines are parallel.*

Given: $\angle 1 = \angle 2$

Prove: $a \parallel b$

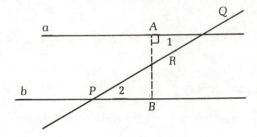

(Analysis: to show that the lines are parallel, we will show they are both perpendicular to the same line $AB$. To do this, we will construct $AB$ perpendicular to one of the lines, and then use congruent triangles to show that it is perpendicular to the other line as well.)

| | |
|---|---|
| 1. Bisect $PQ$, call the midpt. $R$ | 1. Construction |
| 2. Draw $RA \perp a$, extend $AR$ to $B$ | 2. Construction |
| 3. $PR = QR$ | 3. Def. of bisect |
| 4. $\angle PRB = \angle QRA$ | 4. Vertical angles |
| 5. $\angle 1 = \angle 2$ | 5. Given |
| 6. $\triangle PRB \cong \triangle QRA$ | 6. ASA |
| 7. $\angle QAR = \angle PBR$ | 7. CPCTE |
| 8. $\angle QAR = 90°$ | 8. Def. of $\perp$ |
| 9. $\angle PBR = 90°$ | 9. Substitution |
| 10. $RB \perp b$ | 10. Def. of $\perp$ |
| 11. $a \parallel b$ | 11. Two lines $\perp$ to the same line are $\parallel$ |

It is now easy to show, as a corollary to the above theorem, that:

> *If two lines are cut by a transversal so that the corresponding angles are equal, then the lines are parallel.*

Given: ∡1 = ∡2

Prove: a ∥ b

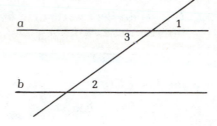

| | |
|---|---|
| **1.** ∡1 = ∡2 | **1.** Given |
| **2.** ∡1 = ∡3 | **2.** Vertical angles are = |
| **3.** ∡2 = ∡3 | **3.** Substitution |
| **4.** a ∥ b | **4.** Alt. int. angles are = |

There are two other corollaries which could be proven but, in use, they only save one or two steps. We mention them here only so that you will recognize them as a situation that will lead to parallel lines: *"If two lines are cut by a transversal so that the alternate exterior angles are equal, then the lines are parallel."* And "If two lines are cut by a transversal so that a pair of interior angles on the same side of the transversal are supplementary, then the lines are parallel."

Here are some examples using the parallel theorems:

**If four shelves are built perpendicular to a vertical line on the wall, will they be parallel to each other?**

**Yes—lines perpendicular to the same line are parallel.**

Another example:

Given: AB = AC
     ∡1 = ∡3

Prove: AD ∥ BC

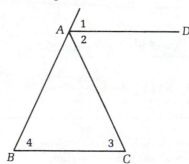

| | |
|---|---|
| **1.** AB = AC | **1.** Given |
| **2.** ∡3 = ∡4 | **2.** ∡s opp. = sides are = |
| **3.** ∡1 = ∡3 | **3.** Given |
| **4.** ∡1 = ∡4 | **4.** Substitution |
| **5.** AD ∥ BC | **5.** Corr. ∡s are = |

We now look at some of the results obtained when we know that a pair of lines are parallel. The theorems and corollaries which follow should be recognized as the converses of the theorems and corollaries we use to prove lines parallel.

---

*If two parallel lines are cut by a transversal, then the alternate interior angles will be equal.*

Given: $AB \parallel CD$

Prove: $\angle ABC = \angle BCD$

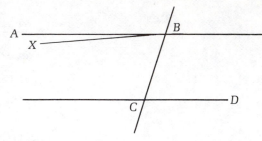

(Analysis: We will use a coincidence proof—that is, we will construct $\angle XBC$ so that it *is* equal to $\angle BCD$, then show that $XB$ and $AB$ are really the same line—so that $\angle XBC$ is also $\angle ABC$, and therefore $\angle ABC = \angle BCD$. Notice that in a coincidence proof we have to draw a distorted figure in order to discuss lines $AB$ and $XB$, when in reality, they will turn out to be the same line.)

| | |
|---|---|
| **1.** Through $B$ construct $\angle XBC = \angle BCD$ | **1.** Construction |
| **2.** $XB \parallel CD$ | **2.** If the alt. int. angles are = the lines are parallel |
| **3.** $AB \parallel CD$ | **3.** Given |
| **4.** $XB$ and $AB$ coincide | **4.** Through a given point only one line can be drawn parallel to a given line. |
| **5.** $\angle ABC = \angle XBC$ | **5.** Same angle (they coincide) |
| **6.** $\angle ABC = \angle BCD$ | **6.** Substitution |

---

And the corollary to this would be:

---

*If two parallel lines are cut by a transversal, then the corresponding angles are equal.*

Given: $a \parallel b$

Prove: $\angle 1 = \angle 2$

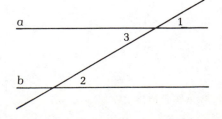

| | |
|---|---|
| **1.** $a \parallel b$ | 1. Given |
| **2.** $\angle 2 = \angle 3$ | 2. Alt. int. angles are = |
| **3.** $\angle 1 = \angle 3$ | 3. Vertical angles |
| **4.** $\angle 1 = \angle 2$ | 4. Substitution |

Some more examples:
**Example 1:**
Given: $AC = BC$
  $DE \parallel AC$

Prove: $DE = BE$

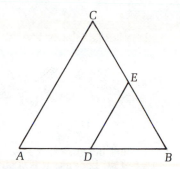

| | |
|---|---|
| **1.** $DE \parallel AC$ | 1. Given |
| **2.** $\angle A = \angle D$ | 2. Corr. $\angle$s are = |
| **3.** $AC = BC$ | 3. Given |
| **4.** $\angle A = \angle B$ | 4. $\angle$s opp. = sides are = |
| **5.** $\angle D = \angle B$ | 5. Substitution |
| **6.** $DE = BE$ | 6. Sides opp. = $\angle$s are = |

**Example 2:**
*Given: $a \parallel b, c \perp d, \angle 1 = 23°$*
**Find all the other numbered angles.**

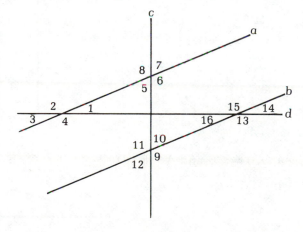

**Using vertical angles, supplementary angles, complementary angles ($\angle 1$ and
$\angle 5$), alternate interior angles, and corresponding angles we get:**

$$\angle 1 = \angle 3 = \angle 14 = \angle 16 = \ \ 23°$$
$$\angle 2 = \angle 4 = \angle 13 = \angle 15 = 157°$$
$$\angle 5 = \angle 7 = \angle 10 = \angle 12 = \ \ 67°$$
$$\angle 6 = \angle 8 = \angle 9 \ \ = \angle 11 = 113°$$

**Can you supply the reason for each equality?**

Example 3:

Two parallel boards are to be supported on a triangular frame as illustrated:

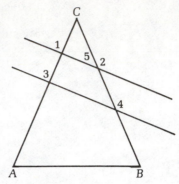

∢C = 64°, ∢1 = 90°, ∢5 = 26°. Find ∢3, ∢2, ∢4.

∢3 = 90° because it is a corresponding angle to ∢1.
∢2 = 154° because it is a supplement to ∢5.
∢4 = 154° because it is a corresponding angle to ∢2.

**EXERCISE SET 12**

1. Given: ∢1 = ∢2

   Prove: $AB \parallel CD$

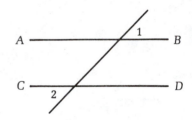

2. Given: ∢1 supplementary to ∢2

   Prove: $AB \parallel CD$

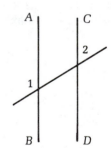

3. Given: $AB = CD$ and $AB \parallel CD$

   Prove: $\triangle ACD \cong \triangle ACB$

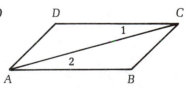

4. Given: $AE = ED$ and $BE = CE$

   Prove: $AB \parallel CD$

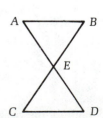

5. Given: $AC \parallel DE$
    $BC \parallel EF$
    $AB = DF$

    Prove: $\triangle ABC \cong \triangle DEF$

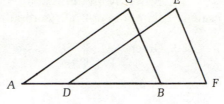

6. $AB \parallel CD$, $\angle EGF = 70°$, $\angle DFG = 135°$. Find the measure of $\angle CEG$

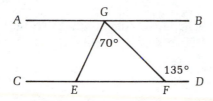

7–10. Using this figure, in which $a \parallel b$, find the indicated angles:

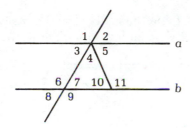

7. $\angle 4 = 45°$, $\angle 10 = 55°$; find $\angle 3$

8. $\angle 5 = 45°$, $\angle 6 = 120°$; find $\angle 4$

9. $\angle 2 + \angle 5 = 125°$, $\angle 11 = 130°$; find $\angle 8$

10. $\angle 1 = 140°$; find $\angle 9$

11–18. Assuming that $a \parallel b$ and $\angle 7 = \frac{1}{5}\angle 1$, find the value of each indicated angle. Hint: Let $\angle 1 = 5X$

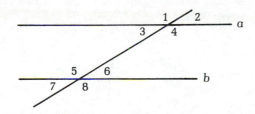

11. $\angle 1 =$          15. $\angle 5 =$

12. $\angle 2 =$          16. $\angle 6 =$

13. $\angle 3 =$          17. $\angle 7 =$

14. $\angle 4 =$          18. $\angle 8 =$

19. Prove by an indirect method that the bisector of an angle of a scalene triangle is not perpendicular to the opposite side.

20. *Given:* $AB \parallel CD$
   $PQ \parallel RS$

   *Prove:* $\angle 1 = \angle 2$

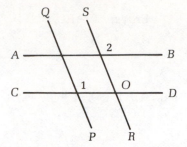

21. *Given:* $\triangle ABC$ is isosceles with $AC = BC$
   $\angle 2$ is constructed equal to $\angle 1$

   *Prove:* $AD \parallel BC$

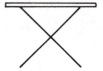

22. Prove that a table constructed so that its supports bisect each other will always be level (parallel to the floor).

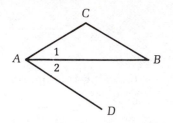

23. *Given:* $PR = QS$
   $PS \parallel QR$
   $QT = RT$

   *Prove:* $PT = ST$

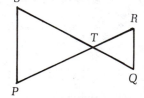

24. Three parallel pipes are to be mounted on a triangular brace. If $\angle 1 = 79°$ and $\angle 4 = 121°$, find angles 2, 3, 5, 6, and the angle at the vertex of the brace.

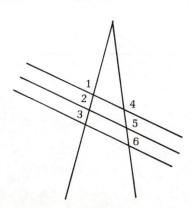

*25. Prove that if two adjacent angles of a four-sided figure are right angles, the bisectors of the other two angles are perpendicular to each other.

*26. Boronoff, Pavlov, Revitsky, and Sukarek are four talented creative artists; one a dancer, one a painter, one a singer, and one a writer (though not necessarily in that order).

   1. Boronoff and Revitsky were in the audience the night the singer made his debut on the concert stage.
   2. Both Pavlov and the writer have sat for portraits by the painter.
   3. The writer, whose biography of Sukarek was a best-seller, is planning to write a biography of Boronoff.
   4. Boronoff has never heard of Revitsky.

What is each man's artistic field?

*27. A recent murder case revolved around six men: Clayton, Forbes, Graham, Holgate, McFee, and Warren. In one order or another, these men were the victim, the murderer, the witness, the policeman, the judge, and the hangman. The facts of the case were simple: the victim had died instantly from the effects of a gunshot wound inflicted at close range. The witness did not see the crime committed, but swore to hearing an argument followed by a shot. After a lengthy trial the murderer was convicted, sentenced to death, and hanged.

   1. McFee knew both the victim and the murderer.
   2. In court the judge asked Clayton to give his account of the shooting.
   3. Warren was the last person to see Forbes alive.
   4. The policeman testified that he picked up Graham near the place where the body was found.
   5. Holgate and Warren never met.

What role did each man play in this unfortunate melodrama? (Some additional assumptions or guesses may be necessary.)

*28. Vernon, Wilson, and Yates are three professional men, one an architect, one a doctor, and one a lawyer, who occupy offices on different floors of the same building. Their secretaries are named, though not necessarily respectively, Miss Ainsley, Ms. Barnette, and Miss Coulter.

   1. The lawyer has his office on the ground floor.
   2. Instead of marrying her boss the way secretaries do in stories, Ms. Barnette plans to marry Yates, and goes out to lunch with him every day.
   3. At noon Miss Ainsley goes upstairs to eat lunch with Wilson's secretary.
   4. Vernon had to send his secretary down to borrow some stamps from the architect's office the other day.

What is each man's profession, and what is the name of each man's secretary? (More than one table may be necessary.)

**\*\*29.** Five men, each of a different nationality, live in different colored houses, drink different drinks, smoke different brands of cigarettes, and own different animals.

1. There are five houses in a row.
2. The Englishman lives in the red house.
3. The Spaniard has a dog.
4. The person who lives in the green house drinks coffee.
5. The Ukrainian drinks tea.
6. The green house is just to the right of the ivory house.
7. The man who smokes Old Golds owns snails.
8. Kools are smoked in the yellow house.
9. The man in the middle house drinks milk.
10. The Norwegian lives in the first house.
11. The man who smokes Kools lives next to the man who owns a horse.
12. The Lucky Strike smoker drinks orange juice.
13. The Japanese smokes Parliaments.
14. The Norwegian lives next to the blue house.
15. The Chesterfield smoker lives next to the man with a fox.

Which man drinks water?
Which man owns a zebra?

**\*\*30.** This problem looks simple, but is very difficult.

*Given:* $AD$ bisects $\sphericalangle BAC$
$\qquad$ $BE$ bisects $\sphericalangle ABC$
$\qquad$ $AD = BE$

*Prove:* $\triangle ABC$ is isosceles

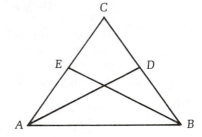

In words: If the angle bisectors of a triangle are equal, then the triangle is isosceles.

---

*3—Summary*

Theorems on congruency and isosceles triangles:

If two triangles have two sides and the included angle of one equal respectively to two sides and the included angle of the other, then they are congruent. SAS

If two right triangles have their legs respectively equal, then the triangles are congruent.

The bisector of the vertex angle of an isosceles triangle is the perpendicular bisector of the base.

If two triangles have two angles and the included side of one equal respectively to two angles and the included side of the other, then they are congruent. ASA

If two right triangles have a leg and adjacent angle respectively equal, then they are congruent.

If two sides of a triangle are equal, then the angles opposite those sides are equal.

If two angles of a triangle are equal, then the sides opposite those angles are equal.

An equilateral triangle is also equiangular.

An equiangular triangle is also equilateral.

If two triangles have three sides of one equal to three sides of the other, then they are congruent. SSS

The seven basic constructions were then proven with the use of the above theorems.

Indirect proofs:

*Reductio ad absurdum:*

When only two possibilities exist, and one can be reduced to an absurdity, then the other must be true.

Exclusion:

When more than two possibilities exist, and we are certain that all possibilities are considered, and can exclude all except one, then the remaining one must be true.

Coincidence:

When two figures can be made to coincide, each will have all the properties of the other.

Theorems on parallel lines:

Two lines in the same plane perpendicular to the same line are parallel.

Two lines in the same plane parallel to the same line are parallel to each other.

If two lines are cut by a transversal so that the alternate interior angles are equal, then the lines are parallel.

If two lines are cut by a transversal so that the corresponding angles are equal, then the lines are parallel.

If two parallel lines are cut by a transversal, then the alternate interior angles will be equal.

If two parallel lines are cut by a transversal, then the corresponding angles will be equal.

Astronomical clock, made by Nicholas Oursian in 1540. Hampton Court Palace, England.
*Photograph by The Bettman Archive.*

# CHAPTER 4

# NUMERICAL RELATIONSHIPS

In this chapter, we consider some of the more functional relationships. Theorems about the size of the angles in a triangle, ratios and proportions of the lengths of sides, similarity of figures, and the most useful Pythagorean Theorem. These are the theorems that are most used in "everyday" or applied problems.

## 13. Angle relationships in a triangle

Now that we have some theorems about parallel lines, we can prove one of the most basic of all the theorems of plane geometry. The theorem which follows, and the Pythagorean Theorem (which was mentioned earlier, but has not yet been proven) are probably the two most important of all the theorems. Both of these were proven by Pythagoras several hundred years before Euclid wrote his "Elements".

---

### The sum of the angles of a triangle is 180°.

Given: △ ABC

Prove: $\angle A + \angle B + \angle C = 180°$

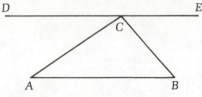

| | |
|---|---|
| **1.** Through C draw DE ∥ AB | **1.** Construction (and Euclidean Parallel Postulate) |
| **2.** $\angle ACD + \angle ACB + \angle BCE = 180°$ | **2.** Def. of st. angle |
| **3.** $\angle ACD = \angle A$ | **3.** Alt. int. angles are = |
| **4.** $\angle BCE = \angle B$ | **4.** (3) |
| **5.** $\angle A + \angle B + \angle C = 180°$ | **5.** Substitution |

---

From this theorem we can now prove a great many other relationships. The five corollaries which follow are some of the more useful ones. First, a definition is necessary. An **exterior angle** of a triangle is formed by one of the sides of the triangle and an adjacent side extended. Every triangle will have six exterior angles:

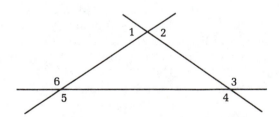

The interior angle of a triangle which is next to a particular exterior angle, is called the **adjacent interior angle,** and the other two interior angles are called the **opposite interior angles.**

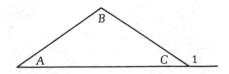

For exterior angle ∡1, the opposite interior angles are ∡A and ∡B. Now for some corollaries:

> *An exterior angle of a triangle is equal to the sum of the opposite interior angles.*

Given: △ABC with AC extended to
      form an exterior angle

Prove: ∡1 = ∡A + ∡B

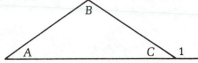

| | |
|---|---|
| **1.** ∡1 + ∡C = 180° | **1.** Def. of st. angle |
| **2.** ∡A + ∡B + ∡C = 180° | **2.** Sum of the angles of a triangle |
| **3.** ∡1 + ∡C = ∡A + ∡B + ∡C | **3.** Substitution |
| **4.** ∡C = ∡C | **4.** Reflexive |
| **5.** ∡1 = ∡A + ∡B | **5.** Subtraction |

As a corollary to the corollary, we can state:

> *An exterior angle of a triangle is greater than either opposite interior angle.*

Given: △ABC with AC extended to
      form an exterior angle

Prove: ∡1 > ∡A and ∡1 > ∡B

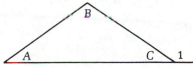

| | |
|---|---|
| **1.** ∡1 = ∡A + ∡B | **1.** Ext. ∡ = sum of opp. int. ∡s |
| **2.** ∡1 > ∡A and ∡1 > ∡B | **2.** The whole is greater than any of its parts. |

*If two angles of one triangle are equal to two angles of another
triangle, then the third angles are equal.*

Given: $\angle A = \angle D$
$\qquad \angle K = \angle G$

Prove: $\angle T = \angle O$

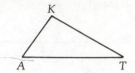

1. $\angle K + \angle A + \angle T = 180°$     1. Sum of angles of a $\triangle$ = 180°

2. $\angle D + \angle O + \angle G = 180°$     2. (1)

3. $\angle K + \angle A + \angle T =$     3. Substitution
$\quad \angle D + \angle O + \angle G$

4. $\angle A = \angle D$     4. Given

5. $\angle K = \angle G$     5. (4)

6. $\angle T = \angle O$     6. Subtraction (of 4. and 5. from 3.)

---

*The acute angles of a right triangle are complementary.*

Given: Right triangle $ABC$ with
right angle at $A$

Prove: $\angle B$ and $\angle C$ are complementary

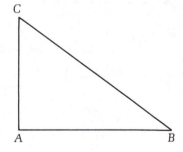

1. $\angle A$ is a right angle     1. Given

2. $\angle A = 90°$     2. Def. of rt. angle

3. $\angle A + \angle B + \angle C = 180°$     3. Sum of angles of a triangle = 180°

4. $\angle B + \angle C = 90°$     4. Subtraction

5. $\angle B$ and $\angle C$ are
complementary     5. Angles whose sum is 90° are complementary

> *Each angle of an equilateral triangle is 60°.*

Given: $\triangle ABC$ is equilateral

Prove: $\angle A = \angle B = \angle C = 60°$

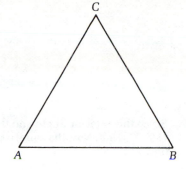

1. $\triangle ABC$ is equilateral

2. $\angle A = \angle B = \angle C$

3. $\angle A + \angle B + \angle C = 180°$

4. $3 \angle A = 180°$

5. $\angle A = 60°$

6. $\angle A = \angle B = \angle C = 60°$

1. Given

2. Equilateral triangles are equiangular

3. Sum of angles of a triangle $= 180°$

4. Substitution

5. Division

6. Substitution

Some examples of problems making use of our new theorems and corollaries:

**Find** $\angle A$

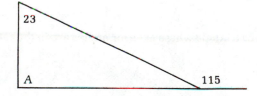

**Since the exterior angle equals the sum of the opposite interior angles, $\angle A +$
$23° = 115°$ therefore $\angle A = 92°$**

In an isosceles triangle, the exterior angle at the vertex is four times as large as the vertex angle. Find the base angles. Drawing a figure always helps:

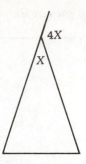

Since the exterior angle and the adjacent angle are supplementary, $X + 4X = 180°, X = 36°$. Now the equal base angles must add to the exterior angle of $4X$, so they are each $2X$, or $72°$.

The acute angles of a right triangle are in the ratio of 2 to 3. How large is the triangle's largest exterior angle?
Draw a figure:

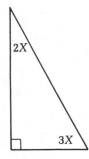

Since the acute angles are complementary, $2X + 3X = 90°$ or $X = 18°$. The angles of the triangle are $36°$, $54°$, and $90°$. Since the largest exterior angle will be adjacent to the smallest interior angle, the largest exterior angle will be supplementary to the $36°$—and therefore $144°$.

Find the angles of a triangle whose exterior angles are in the ratio of 3 to 4 to 5.

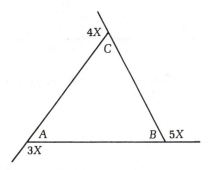

$$5X = \angle A + \angle C$$
$$4X = \angle A + \angle B$$
$$3X = \angle B + \angle C$$
So $12X = 2\angle A + 2\angle B + 2\angle C$
$12X = 360$ (2 times the sum of the angles of a triangle)
$X = 30°, 3X = 90°, 4X = 120°, 5X = 150°$
So the angles of the triangle are $90°$, $60°$, and $30°$.

Given:  $\angle A = \angle D$
   $\angle C = \angle F$
   $BC = EF$

Prove:  $\triangle ABC \cong \triangle DEF$

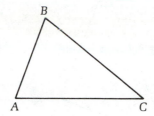 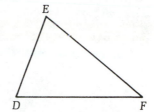

Notice that the side is not the included side, so we cannot use SAS. The plan would be to either show $AC = DF$ and use SAS or show $\angle B = \angle E$ and use ASA. Since there is no information about $AC$ and $DF$, we will use a recent corollary to show $\angle B = \angle E$, and then use ASA.

| | |
|---|---|
| **1.** $\angle A = \angle D$ | **1.** Given |
| **2.** $\angle C = \angle F$ | **2.** Given |
| **3.** $\angle B = \angle E$ | **3.** If 2 angles are equal, the third angles are equal |
| **4.** $BC = EF$ | **4.** Given |
| **5.** $\triangle ABC \cong \triangle DEF$ | **5.** ASA |

(This exercise shows that triangles are congruent if we have AAS = AAS.)

**Show that the altitudes to the equal legs of an isosceles triangle are equal. An altitude is a line from a vertex perpendicular to the opposite side.**

Given:  $AC = BC$
$BD \perp AC$
$AE \perp BC$

Prove:  $AE = BD$

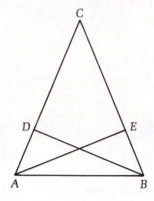

| 1. $BD \perp AC$ and $AE \perp BC$ | 1. Given |
|---|---|
| 2. $\angle BDC = \angle AEC$ | 2. $\perp$ s form rt. angles; all rt. angles are equal |
| 3. $\angle C = \angle C$ | 3. Reflexive |
| 4. $\angle CAE = \angle CBD$ | 4. In triangles with 2 angles equal the third angles are equal |
| 5. $AC = BC$ | 5. Given |
| 6. $\triangle AEC \cong \triangle BDC$ | 6. ASA |
| 7. $AE = BD$ | 7. CPCTE |

Can you prove the converse of this statement: that is, if two altitudes of a triangle are equal, then the triangle is isosceles?

1. Given: *AE* and *BD* are altitudes of △ *ABC*

   Prove: ∢1 = ∢2

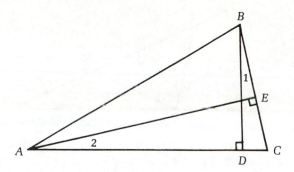

2. Given: ∢ *A* + ∢ *B* + ∢ *D* = 180°

   Prove: *AC* ∥ *DE*

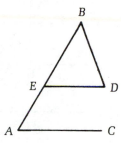

3. Given: △ *ABC* so that ∢1 + ∢2 = 270°

   Prove: △ *ABC* is a right △

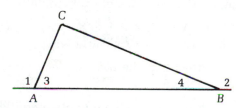

4. Given isosceles triangle *ABC* with *AB* = *AC*
   a)  If ∢ *A* = 32°, find ∢ *C*
   b)  If ∢ *C* = 52°, find ∢1
   c)  If ∢ *A* = 48°, find ∢2

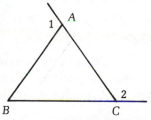

5.  In a right triangle one of the acute angles is 37° 49′ 17″. How large is the other acute angle?

6.  If the angles of a triangle are in the continued ratio of 1:2:3, find the size of the angles. Hint: Let the angles be *X*, 2 *X*, and 3 *X*.

7.  If the first angle of a triangle is twice the second, and the third angle is three times the first, find the angles of the triangle.

**133**

8. Given that $DB = DC = BE$, $\angle EBD = 80°$, find the number of degrees in $\angle ABE$.

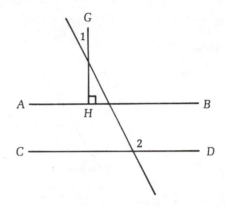

9. Given that $AB \parallel CD$, $GH \perp AB$, $\angle 1 = 25°$ find the number of degrees in $\angle 2$.

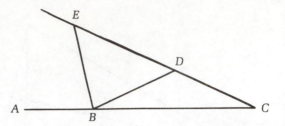

10. The vertex angle of an isosceles triangle is 30° larger than a base angle. Find the size of each angle.

11. In $\triangle ABC$, $\angle A$ is $9X$, $\angle B$ is $3X - 6$, and $\angle C$ is $11X + 2$. Show that $\triangle ABC$ is a right triangle.

12. Prove that the base angles of an isosceles right triangle are each 45°.

13. Prove that each angle of an equilateral triangle is 60°.

14. The exterior angle at the vertex of an isosceles triangle measures $3X + 12$ degrees. One of the opposite interior angles is 30°. Find the measure of the vertex angle.

15. In $\triangle ABC$, $\angle A = 3X$, $\angle B = 5X + 7$, $\angle C = 8X - 3$. Find the angles of the triangle.

16. *Prove:* If two angles and one side (not the included one) of one triangle are equal respectively to two angles and one side of another triangle, then the triangles are congruent. (AAS = AAS)

17. Find the measure of $\angle 1$.

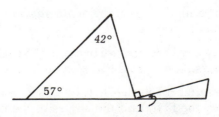

18. Prove $AB \parallel CD$

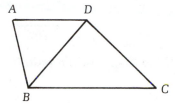

$\angle D = 2(3X + 53)$
$\angle A = 74 - 6X$

19. *Given:* $BD$ bisects $\angle ABC$
$AB = AD$

*Prove:* $AD \parallel BC$

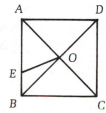

*20. *Given:* $ABCD$ is a square with
diagonals intersecting at O
$AO = AE$

*Prove:* $\angle AOE = 3 \angle BOE$

21. Prove: If two right triangles have a leg and the non-adjacent angle respectively equal, then they are congruent.

22. Prove: If two right triangles have the hypotenuse and an angle respectively equal, then they are congruent.

23. Prove: If two of the altitudes of a triangle are equal, then the triangle is isosceles.

*24. Considering that every triangle has six parts (three sides and three angles), do you think it is possible for two triangles to have five parts of one equal to five parts of the other and *not* be congruent? Either prove that they must be congruent or present a counterexample. (That is, construct two triangles which have five parts equal and are not congruent.)

Some review of the algebraic concepts of ratio and proportion will be useful before we discuss the concept of similarity.

A **ratio** is the comparison of two numbers by division: $\frac{a}{b}$. It is some-times written with a colon: $a{:}b$, and read "the ratio of $a$ to $b$." A ratio of the amount of flour to water in a recipe or the gear ratio in a differential, does not tell how much flour or how many teeth on a gear, but only gives a comparison of the two quantities. Three cups of flour to one cup of water, or 6 tablespoons of flour to 2 tablespoons of water, or 30 grams of

**14. Ratio and
proportion**

flour to 10 grams of water are all a ratio of 3 to 1. In these two triangles, the ratio of the measures of the corresponding sides is 2:1.

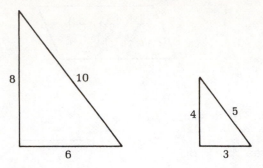

In these triangles, the ratio is 3:5

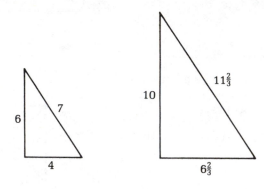

$$\frac{4}{6\frac{2}{3}} = \frac{\frac{12}{3}}{\frac{20}{3}} = \frac{3}{5} \qquad \frac{6}{10} = \frac{3}{5} \qquad \frac{7}{11\frac{2}{3}} = \frac{\frac{21}{3}}{\frac{35}{3}} = \frac{3}{5}$$

The equality of two ratios, $\frac{a}{b} = \frac{c}{d}$, is called a **proportion.** In the proportion $\frac{a}{b} = \frac{c}{d}$, $a$ and $d$ are called the **extremes** and $b$ and $c$ are called the **means.**

We will find the following algebraic rules useful:

$$\frac{a}{b} = \frac{c}{d} \text{ if and only if } ad = bc$$

In words: **The product of the extremes is equal to the product of the means.** This is useful for finding a corresponding part when the ratio is known. For example, if the ratio of the base of a triangle to its height is known to be 3:5, we can find the height of a triangle with base 9 by:

$$\frac{3}{5} = \frac{9}{X}, \ 3X = 45, \ X = 15.$$

**A medium pizza serves 3 people and costs $2.25. If a large pizza serves 5, what should be its proportional cost?**

$$\frac{3}{2.25} = \frac{5}{X}, \ 3X = 5(2.25), \ X = 5(.73), \ X = \$3.65$$

When one number is used for both of the means it is called the **mean proportional** (singular). In $\frac{a}{b} = \frac{b}{c}$, $b$ is the mean proportional to $a$ and $c$. We will see this mean proportional in several of the sections which follow. To find the mean proportional to 9 and 16: $\frac{9}{X} = \frac{X}{16}$, $X^2 = 144$, $X = \pm 12$.

**What is the mean proportional to 3 and $65\frac{1}{3}$?**

$$\frac{3}{X} = \frac{X}{65\frac{1}{3}}, X^2 = 196, X = \pm 14.$$

Also in a proportion, $\frac{a}{b} = \frac{c}{d}$, the terms are numbered: $a$ is the first term, $b$ the second, $c$ the third, and $d$ the fourth term of the proportion. Find the fourth proportional to 3, 5, and 11:

$$\frac{3}{5} = \frac{11}{X}, 3X = 55, X = 18\frac{1}{3}.$$

**If 12 liters of gas in my car will take me 90 km, how far can I go on a full tank if the tank holds 100 liters?**

$$\frac{12}{90} = \frac{100}{X}, 12X = 90(100), X = 750 \text{ km.}$$

When we have a mean proportional, the second and third terms are the same and the fourth term becomes the third. A problem is stated in the following manner:

**Find the third proportional to 3 and 8.**

$$\frac{3}{8} = \frac{8}{X}, 3X = 64, X = 21\frac{1}{3}$$

**Find the third proportional to 5 and 16.**

$$\frac{5}{16} = \frac{16}{X}, 5X = 256, X = 51\frac{1}{5}.$$

Other easily proven properties of proportions are:

If $\frac{a}{b} = \frac{c}{d}$ then $\frac{a}{c} = \frac{b}{d}$          (called alternation)

If $\frac{a}{b} = \frac{c}{d}$ then $\frac{b}{a} = \frac{d}{c}$          (inversion)

If $\frac{a}{b} = \frac{c}{d}$ then $\frac{a+b}{b} = \frac{c+d}{d}$          (addition)

If $\frac{a}{b} = \frac{c}{d}$ and $\frac{a}{b} = \frac{c}{e}$ then $d = e$      (If three terms of one proportion are equal respectively to three terms of another proportion, then the fourth terms are equal.)

1. Find the fourth proportional to 2, 5, and 6 in that order.

2. Find the fourth proportional to 15, 5, and 9 in that order.

3. Find the mean proportional between 16 and 25.

4. Find the mean proportional between 9 and 4.

5. Find the third proportional to 3 and 15.

6. Find the third proportional to 7 and 11.

7. Show that the proportion $\frac{11}{13} = \frac{77}{91}$ is true, by showing that the product of the means is equal to the product of the extremes.

8. *Solve:* $\frac{X}{X + 3} = \frac{4}{5}$

9. *Solve:* $\frac{X}{6} = \frac{5}{12}$

10. *Solve:* $\frac{9}{17} = \frac{X}{85}$

11. Does $\frac{13}{17} = \frac{91}{119}$?

12. Does $\frac{23}{42} = \frac{138}{252}$?

13. Does $\frac{11}{31} = \frac{143}{401}$?

14–18. Name the property that justifies each of the following transformations:

14. If $\frac{a}{b} = \frac{5}{7}$ then $7a = 5b$

15. If $\frac{a}{b} = \frac{5}{7}$ then $\frac{a + b}{b} = \frac{12}{7}$

16. If $\frac{a}{b} = \frac{5}{7}$ then $\frac{a}{5} = \frac{b}{7}$

17. If $\frac{a}{b} = \frac{5}{7}$ then $\frac{b}{a} = \frac{7}{5}$

18. If $\frac{a}{b} = \frac{5}{7}$ and $\frac{a}{b} = \frac{5}{X}$ then $X = 7$

19. A photographer has a slide of which he is very proud, and would like an enlargement for his wall. The wall can only handle a width of 60 cm. If the slide has a width of 3.5 cm and a height of 2.5 cm, what will be the maximum dimensions of the enlargement?

20. A dead tree needs to be felled, but the only direction for it to fall is toward a house which is 20 m away. At a time of day when a meter stick placed perpendicular to the ground casts a shadow of 73 cm, the tree has a shadow of 14 m. If the tree is felled, will it hit the house?

21. Grandma has just received a picture of her son and three grandchildren, Al, Ben, and Chuck. The grandchildren have grown

considerably since she saw them last, and she wants to know their height. Her son, whom she knows to be 5′10″, measures 5.4 cm in the picture. Al, Ben, and Chuck measure 5.6, 5.1, and 4.6 cm. respectively. How tall are the grandchildren?

22. A recipe for brownies calls for a 9 × 13 pan and the only pan available is 9 × 9. What ratio should be used to reduce the recipe?

23.

---

*Mother's Chocolate Cake*

Combine ingredients in order, pour into a greased and floured 9 × 9 pan, and bake 30 min. at 350°.

¾ c. sugar

¼ c. butter

1 egg

¼ c. milk

¼ c. cocoa

1 c. flour

½ c. hot water with 1 tsp. soda added last

---

Since we do not have a 9 × 9 pan (81 sq. in.), we wish to increase the recipe to fill a 9 × 13 pan (117 sq. in.). The ratio should be 81 to 117, but 80 to 120, or 2 to 3 will be close enough. Make the new list of ingredients. (This is a good recipe! If you can't do the assigned problems, try turning in a cake.)

24. Reg can saw a log in half in 8 minutes. How long will it take him to saw it into three equal pieces?

*25. A child's gym set has the following dimensions:

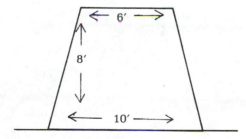

A chinning bar is to be placed between the two legs at a height of 5′. How long does the bar need to be?

*26. If it would take a cannon ball 3⅜ seconds to travel 4 miles, and 3⅜ seconds to travel the next 4, and 3⅜ seconds to travel the next four, and if the rate of progress continues to diminish in the same ratio, how long would it take to go fifteen hundred million miles?—*Arithmeticus*

I don't know.—*Mark Twain*

Can you do any better than Mark Twain on this problem?

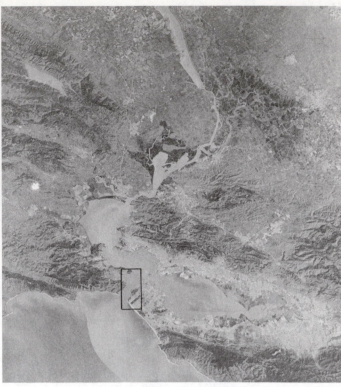

The picture of the San Francisco Bay area on the right was taken from Skylab at an altitude of 435 km. The lower altitude photo, on the left, was taken from a U-2 plane at an altitude of 20 km. This is an example of similarity. *National Aeronautics and Space Administration.*

## 15. Similarity

As we have mentioned previously, the concept of similarity (symbol ~) is that of figures having the "same shape". When is it that figures will have the same shape? Of course, congruent figures will be similar, but we are interested here in what conditions will make non-congruent figures similar. Two 30–60–90 triangles have the same shape even if their sides are not respectively equal.

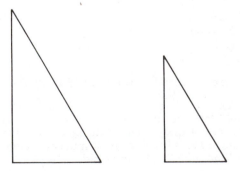

So having the angles of one figure equal to the angles of another seems to work for making triangles similar. Does it also work for figures of more than three sides? Would you call a square and a rectangle similar?

What about figures with no angles? Are any two circles similar?

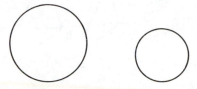

We are going to need more than our intuition to handle this concept of similarity. Once again we can use some of the concepts of transformations to help us.

We will define **similarity** as a **transformation which preserves ratios of distances.** The types of transformations which we will use will also preserve angle measures, collinearity, and betweeness. If the ratios of distances between corresponding points are proportional we will have a constant of proportionality, $K$, such that $AB = K(A'B')$. $K$ is sometimes called the ratio of similitude. The easiest such transformation to consider is called a **dilatation.** It can be thought of as an object and its shadow. A dilatation of a figure composed of line segments **will transform each line into a parallel line.** Point $P$ is called the center of similitude.

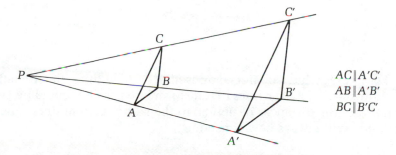

$AC \parallel A'C'$
$AB \parallel A'B'$
$BC \parallel B'C'$

Another type of transformation which yields similar figures is called a **half-turn.** It also transforms each line into a parallel line, except in this case the center of similitude, $P$, is between the two figures.

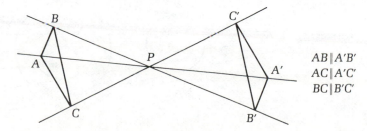

$AB \parallel A'B'$
$AC \parallel A'C'$
$BC \parallel B'C'$

In each of these transformations the ratio of similitude is the ratio of the distances $PA:PA'$.

If we wish a triangle half again as large as $\triangle ABC$, we can make $\dfrac{PA}{PA'} = \dfrac{2}{3}$. Draw rays through each vertex and construct $BA \parallel B'A'$ and $AC \parallel A'C'$

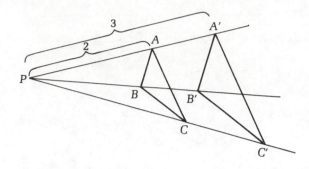

Or, using a half-turn:

$$\frac{PA}{PA'} = \frac{2}{3}$$

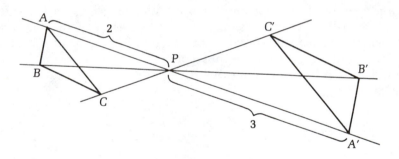

For figures which are not composed of straight lines, the dilatation or half-turn will not have lines to transform into parallel lines; however, the ratios of the distances of points, to their images from the center of similitude will always be a constant.

$$\frac{PA}{PA'} = \frac{PB}{PB'} = \frac{PC}{PC'} = \text{etc.} = K$$

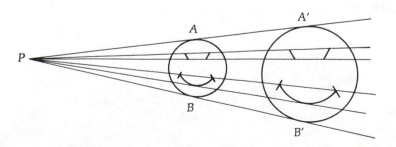

It is interesting to note that any two circles will be similar, and they will have two centers of similitude, $P_1$ and $P_2$. One circle can be either a dilatation or a half-turn of the other.

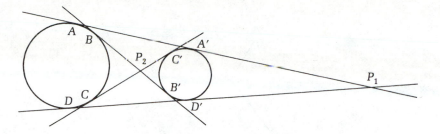

With careful measurement (by all means, use metric) and a little arithmetic, you should be able to verify that the ratio of similitude is the same for both the dilatation and the half-turn. That is,

$$P_1A' : P_1A = P_2C' : P_2C.$$

Note that figures which are formed by a dilatation can overlap:

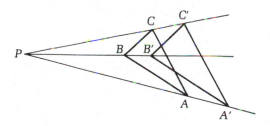

They may also have one of their points as the center of similitude:

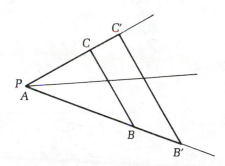

In the illustrations which follow, a dilatation and a half-turn, check to verify that lines are transformed into parallel lines, angle measure is preserved, and the ratios of any two corresponding sides is the constant $K = \dfrac{PA'}{PA}$. Measure the angles with a protractor. Measure the line segments to the nearest millimeter. (Realize that measurement never constitutes a proof—as all measurements are inaccurate.)

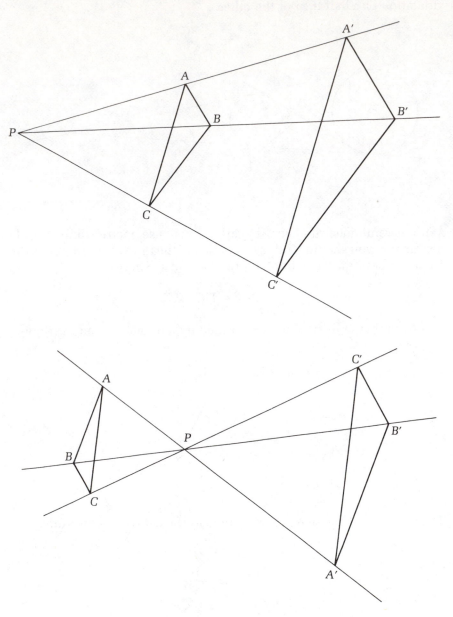

A pantograph is a mechanical device for reproducing a map, drawing or other figure on a larger or smaller scale.

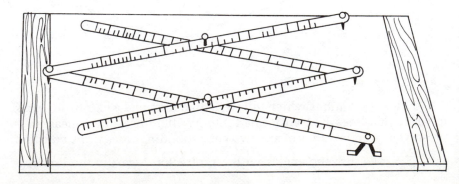

It joins the rods in such a manner so as to maintain a constant ratio between the original figure and the new one. Thus one figure is a dilatation of the other and the two figures are similar. A poor-man's pantograph can be made from a rubber band with a knot in it. Since the sections of the rubber band on either side of the knot each stretch proportionally (until the band breaks), the ratios of distances will form a dilatation and the figures traced will be similar.

Anchor one end of the rubber band to the paper, put a pencil in the loop at the other end, and move the pencil so that the knot will trace the original figure.

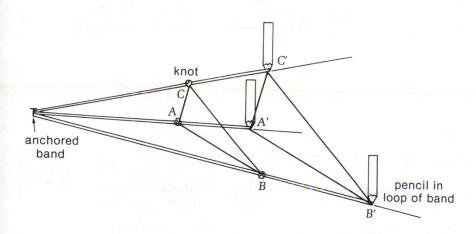

The ratio of similitude can be changed by changing the position of the anchored end and the position of the knot in the rubber band. Try tracing a figure by this method.

Since it is usually not convenient to prove two figures similar by demonstrating that one is a particular type of transformation of the other, we need to develop some theorems regarding similarity.

*A line through two sides of a triangle, parallel to the third side, cuts off a similar triangle.*

Given: DE ∥ AB

Prove: △DEC ∼ △ABC

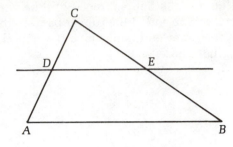

| | |
|---|---|
| **1.** AC ∥ CD | **1.** Same line (It is proper to consider a line as being parallel to itself) |
| **2.** CE ∥ BC | **2.** Same line |
| **3.** DE ∥ AB | **3.** Given |
| **4.** △CDE is a dilatation of △ABC | **4.** Def. of dilatation (C is the center of similitude) |
| **5.** △DEC ∼ △ABC | **5.** Def. of similar (One triangle is a dilatation of the other) |

As a corollary to this theorem, we have what is one of the most useful statements about similar triangles.

*A line through two sides of a triangle, parallel to the third side, divides the sides proportionally.*

Given: DE ∥ AB

Prove: $\dfrac{CD}{AC} = \dfrac{CE}{BC}$

Use the previous figure.

| | |
|---|---|
| **1.** DE ∥ AB | **1.** Given |
| **2.** △CDE ∼ △ABC | **2.** ∥ lines cuts off a ∼△ |
| **3.** $\dfrac{CD}{AC} = \dfrac{CE}{BC}$ | **3.** Similarity is a transformation which preserves proportionality of line segments. |

Now that we have the corollary about proportional segments in a triangle, we can extend it to:

---

*Three or more parallel lines cut off proportional segments on any two transversals.*

Given: $AA' \parallel BB' \parallel CC' \parallel DD'$

Prove: $\dfrac{AB}{A'B'} = \dfrac{BC}{B'C'} = \dfrac{CD}{C'D'}$

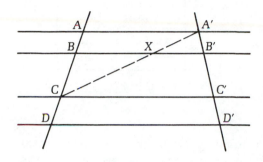

1. Draw $CA'$

   1. Construction

2. $BB' \parallel AA'$

   2. Given

3. In $\triangle ACA'$, $\dfrac{AB}{BC} = \dfrac{A'X}{CX}$

   3. $\parallel$ line divides the sides proportionally

4. In $\triangle A'CC'$, $\dfrac{A'X}{CX} = \dfrac{A'B'}{B'C'}$

   4. $\parallel$ line divides the sides proportionally

5. $\dfrac{AB}{BC} = \dfrac{A'B'}{B'C'}$

   5. Substitution

6. $\dfrac{AB}{A'B'} = \dfrac{BC}{B'C'}$

   6. Alternation

In like manner, draw $DB'$ and continue for proportionality of the other segments.

This corollary gives us a convenient method of dividing *any* line segment (board, piece of cloth, and so on) into *any* number of equal parts. If some equally-spaced parallel lines are available, such as lined notebook paper, simply lay the object diagonally across the lines, so that it intersects the desired number of lines. For example, to divide a nine-inch line into seven equal parts, it should be placed so that it intersects seven spaces.

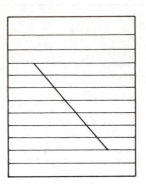

Isn't this considerably easier than trying to measure sections of $1\frac{2}{7}$ inches? Even with metric measures, this method is easier. If no convenient-sized grid of equally spaced parallel lines is available, we can proceed this way:

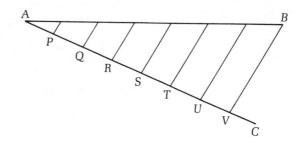

**To divide $AB$ into equal segments—**
    **Draw any other line $AC$**
    **Mark off, on $AC$, $n$ equal segments of any convenient length (for illustration we will use $n = 7$)**
    **Join the last segment to $B$**
    **Draw lines parallel to $BV$ through each of the other division points**
    **These will divide $AB$ into n equal parts.**
    **(Since the ratios on $AC$ are 1:1, the ratios on transversal $AB$ are 1:1.)**

A very simple and convenient method of proving a pair of triangles similar is given by the theorem:

> *If two angles of one triangle are equal to two angles of another triangle, then the triangles are similar.* **(Abbreviation—AA, sim.)**

Given: $\angle A = \angle A'$, $\angle C = \angle C'$

Prove: $\triangle A'B'C' \sim \triangle ABC$

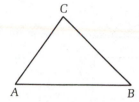 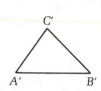 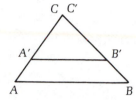

1. Translate (or reflect & translate) $A'B'C'$ so that $C'$ coincides with $C$ and $A'C'$ falls on $AC$

2. $\angle C = \angle C'$

3. $C'B'$ falls on $CB$

4. $\angle A = \angle A'$

5. $A'B' \parallel AB$

6. $\triangle A'B'C' \sim \triangle ABC$

 

1. Existence of translation of any line to another

2. Given

3. $\angle$ measure postulate

4. Given

5. = corresponding angles

6. $\parallel$ line cuts off $\sim \triangle$

Before we can prove our next method of showing triangles similar, we need the following theorem. Notice that this theorem is the converse of a previous theorem.

---

*If a line divides two sides of a triangle proportionally, then it is parallel to the third side.*

---

Given: $\dfrac{AC}{CD} = \dfrac{BC}{CE}$

Prove: $DE \parallel AB$

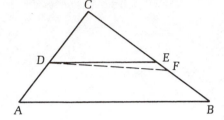

This will be an indirect proof, using the method of coincidence. We will show that $DE$ and $DF$ coincide, so that each possesses the properties of the other.

| | |
|---|---|
| **1.** Draw $DF \parallel AB$ | 1. Construction |
| **2.** $\dfrac{AC}{CD} = \dfrac{BC}{CF}$ | 2. $\parallel$ line divides proportionally |
| **3.** $\dfrac{AC}{CD} = \dfrac{BC}{CE}$ | 3. Given |
| **4.** $CF = CE$ | 4. Substitution (One of our properties of proportions) |
| **5.** $E$ and $F$ coincide | 5. Line Measure Postulate |
| **6.** $DE$ and $DF$ coincide | 6. Two points determine a line |
| **7.** $DE \parallel AB$ | 7. Each line has the properties of the other (Method of coincidence) |

With the use of this theorem we can now derive a third method of proving triangles similar.

> *If two triangles have an angle of one equal to an an angle of the*
> *other and the including sides proportional, then the triangles*
> *are similar.* **(SAS, sim.)**

Given: $\angle C = \angle F$, $\dfrac{AC}{DF} = \dfrac{BC}{EF}$

Prove: $\triangle ABC \sim \triangle DEF$

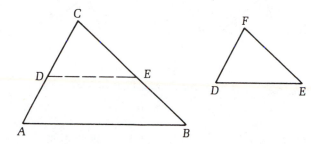

1. Translate (or rotate and translate) $\triangle DEF$ so that $F$ coincides with $C$ and $DF$ falls on $AC$

   1. Existence of a translation

2. $\angle C = \angle F$

   2. Given

3. $EF$ lies on $BC$

   3. Angle Measure Postulate

4. $\dfrac{AC}{DF} = \dfrac{BC}{EF}$

   4. Given

5. $DE \parallel AB$

   5. Line dividing two sides prop. is $\parallel$ to the third side.

6. $\triangle ABC \sim \triangle DEF$

   6. $\parallel$ line cuts off a $\sim \triangle$

In the proof of the fourth method of showing triangles similar, we will use small letters to designate line segments. This will simplify our writing of several proportions. Notice also that congruent triangles are similar.

> *If two triangles have their sides respectively proportional, then they are similar.* **(SSS, sim.)**

Given: $\dfrac{a}{a'} = \dfrac{b}{b'} = \dfrac{c}{c'}$

Prove: $\triangle ABC \sim \triangle A'B'C'$

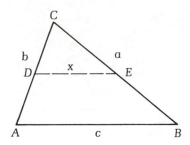
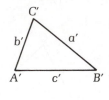

| | |
|---|---|
| **1.** Mark off $CD = C'A'$ | 1. Construction |
| **2.** Mark off $CE = C'B'$ | 2. Construction |
| **3.** Draw $DE$, and call it x | 3. Construction |
| **4.** $\dfrac{a}{a'} = \dfrac{b}{b'}$ | 4. Given |
| **5.** $\angle C = \angle C$ | 5. Reflexive |
| **6.** $\triangle CDE \sim \triangle ABC$ | 6. SAS, sim. |
| **7.** $\dfrac{a}{a'} = \dfrac{c}{x}$ | 7. Corr. sides of $\sim \triangle$s |
| **8.** $\dfrac{a}{a'} = \dfrac{c}{c'}$ | 8. Given |
| **9.** $x = c'$ | 9. 3 terms of 2 proportions $=$, the 4th are $=$ |
| **10.** $\triangle CDE \cong \triangle A'B'C'$ | 10. SSS |
| **11.** $\triangle ABC \sim \triangle A'B'C'$ | 11. Substitution ("A quant. may be substituted for its equal in an expression" applies to similarity expressions.) |

Many of the theorems we have proven so far were ones you would have assumed as true by using good sense or intuition. The theorem which follows is a surprising one. Students rarely suspect that this useful and interesting relationship exists when an angle is bisected. In the proof we make use of several auxiliary lines.

> *The bisector of an angle of a triangle divides the opposite side into segments proportional to their adjacent sides.*

Given: CD bisects ∢C

Prove: $\dfrac{AD}{BD} = \dfrac{AC}{BC}$

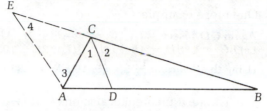

| | |
|---|---|
| **1.** Draw $AE \parallel CD$ | **1.** Construction |
| **2.** Extend $BC$ to intersect $AE$ | **2.** Construction |
| **3.** $CD$ bisects ∢$C$ | **3.** Given |
| **4.** ∢1 = ∢2 | **4.** Def. of bisect |
| **5.** ∢1 = ∢3 | **5.** Alt. int. angles |
| **6.** ∢2 = ∢4 | **6.** Corr. angles |
| **7.** ∢3 = ∢4 | **7.** Quant. = to = quant. are = to each other. |
| **8.** $AC = CE$ | **8.** Sides opp. = angles |
| **9.** In $\triangle ABE$, $\dfrac{AD}{BD} = \dfrac{CE}{BC}$ | **9.** $\parallel$ line divides sides prop. |
| **10.** $\dfrac{AD}{BD} = \dfrac{AC}{BC}$ | **10.** Substitution |

Let us illustrate the use of this theorem with some numerical examples:

In $\triangle ABC$, **CD** bisects ∢$C$, $AC = 6$, $BC = 9$, $AB = 12$. Find $AD$ and $BD$. Let $AD = X$, then $BD = 12 - X$

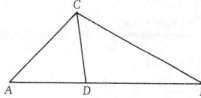

By the theorem: $\dfrac{AD}{BD} = \dfrac{AC}{BC}$, so

$\dfrac{X}{12 - X} = \dfrac{6}{9}$, $9X = 72 - 6X$

$15X = 72$, $X = 4.8 = AD$, $BD = 12 - X = 7.2$

NOTE that this theorem gives us a practical way of bisecting an angle without the use of a compass. By using the proportion to find the length of $AD$, we can measure to locate $D$. We then draw the line $CD$, which will be the bisector of ∢$ACB$!

Using the same figure, $CD$ bisects $\angle C$, $AC = 8$, $BC = 10$, $AD = 6\frac{2}{3}$. Find $AB$.

Let $BD = X$. $\dfrac{\frac{20}{3}}{X} = \dfrac{8}{10}$, $8X = \dfrac{200}{3}$, $BD = X = 8\frac{1}{3}$, $AB = AD + BD$, $AB = 6\frac{2}{3} + 8\frac{1}{3} = 15$

One more example:

**Again** $CD$ bisects $\angle C$, $AC = 12$, $AB = 28$, $AD = 10.5$. Find $BC$.
Let $BC = X$, $BD = AB - AD$, $BD = 28 - 10.5 = 17.5$.
**Using the theorem:** $\dfrac{10.5}{17.5} = \dfrac{12}{X}$, $10.5\,X = 12(17.5)$, $BC = X = 20$

The next theorem is also one that is not intuitively too evident. It is also very important to us, as it is the key theorem in our proof of the very important Pythagorean Theorem. (It also makes nice numerical problems for tests!)

---

> *The altitude to the hypotenuse of a right triangle forms three similar triangles.*

Recall that the sides adjacent to the right angle are called legs, and the side opposite the right angle is called the hypotenuse.

Given: $\angle ACB$ is a right angle,
$\quad\quad\quad CD \perp AB$

Prove: $\triangle ABC \sim \triangle ACD \sim \triangle BCD$

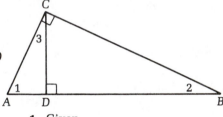

| | |
|---|---|
| 1. $\angle ACB$ is a right angle | 1. Given |
| 2. $\angle 1$ and $\angle 2$ are complementary | 2. Def. of comp. |
| 3. $CD \perp AB$ | 3. Given |
| 4. $\angle ADC$ is a right triangle | 4. Def. of rt. triangle |
| 5. $\angle 1$ and $\angle 3$ are complementary | 5. Acute angles of a rt. triangle |
| 6. $\angle 2 = \angle 3$ | 6. Comp. of the same angle |
| 7. $\angle ADC = \angle BDC$ | 7. All rt. angles are = |
| 8. $\triangle ACD \sim \triangle BCD$ | 8. AA, sim. |
| 9. $\angle ADC = \angle ACB$ | 9. All rt. angles are = |
| 10. $\angle 3 = \angle 3$ | 10. Reflexive |
| 11. $\triangle ACD \sim \triangle ABC$ | 11. AA, sim. |
| 12. $\triangle ABC \sim \triangle ACD \sim \triangle BCD$ | 12. Substitution |

Before illustrating the use of this theorem with some numerical examples we should first prove one quick corollary.

**155**

**15–SIMILARITY**

*The altitude to the hypotenuse of a right triangle is the mean proportional to the segments of the hypotenuse.*

Given: $\angle ACB$ is a right angle,
$\quad\quad CD \perp AB$

Prove: $\dfrac{AD}{CD} = \dfrac{CD}{BD}$

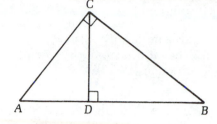

| | |
|---|---|
| 1. $\angle ACB$ is a right angle | 1. Given |
| 2. $CD \perp AB$ | 2. Given |
| 3. $\triangle CDA \sim \triangle BDC$ | 3. Alt to hyp. of rt. $\triangle$ makes $\sim \triangle$s |
| 4. $\dfrac{AD}{CD} = \dfrac{CD}{BD}$ | 4. Corr. parts of similar $\triangle$s |

In the examples which follow, use this figure in which $\angle ACB$ is a right angle, $CD \perp AB$.

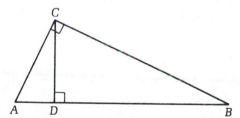

If $AD = 5, BD = 8$, find $CD$.

Let $CD = X$ then by the theorem $\dfrac{AD}{CD} = \dfrac{CD}{BD}$ or $\dfrac{5}{X} = \dfrac{X}{8}, X^2 = 40, CD = X = 2\sqrt{10}$
($-2\sqrt{10}$ also solves the equation, but our line segments do not have direction.)

If $AD = 5, CD = 10$, find $BD$

Let $BD = X$ then $\dfrac{5}{10} = \dfrac{10}{X}, 5X = 100, BD = X = 20$

If $BD = 27, CD = 9$, find $AB$

Let $AB = X, AD = X - 27$ then $\dfrac{X - 27}{9} = \dfrac{9}{27}, 27(X - 27) = 81, 27X = 810,$
$AB = X = 30$

If $AD = 3, CD = 4, AC = 5$, find $BC$ (use similar triangles)

Let $BC = X$ then $\dfrac{3}{5} = \dfrac{4}{X}, 3X = 20, BC = X = 6\dfrac{2}{3}$

Since quite a few theorems and corollaries were presented in this section, it might be helpful to summarize.

**METHODS OF PROVING TRIANGLES SIMILAR:**

1. A line through two sides of a triangle, parallel to the third side, cuts off a similar triangle.
2. If two angles of one triangle are equal to two angles of another triangle, then the triangles are similar. (AA, sim.)
3. If two triangles have an angle of one equal to an angle of the other and the including sides proportional, then the triangles are similar. (SAS, sim.)
4. If two triangles have their sides respectively proportional, then they are similar. (SSS, sim.)
5. The altitude to the hypotenuse of a right triangle forms three similar triangles.

**THEOREMS ABOUT PROPORTIONS:**

1. A line through two sides of a triangle, parallel to the third side, divides the sides proportionally. And its converse:
2. If a line divides two sides of a triangle proportionally, then it is parallel to the third side. (This is a new method to show ‖ .)
3. Three or more parallel lines cut off proportional segments on any two transversals.
4. The bisector of an angle of a triangle divides the opposite side into segments proportional to their adjacent sides.
5. The altitude to the hypotenuse of a right triangle is the mean proportional to the segments of the hypotenuse.

1–4. Refer to this figure in which $DE \parallel AB$

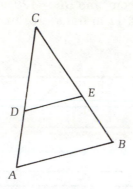

1. If $EB = 2$, $EC = 5$, and $AD = 3$, find $CD$.

2. If $AC = 12$, $BC = 20$, and $EB = 15$, find $AD$ and $DC$.

3. If $AD = \frac{1}{2}CD$, find the ratio of $CE$ to $EB$.

4. If $AC = 10$, $AD = 2$, and $EB = 3$, find $BC$ and $CE$.

5. In this figure, $BD$ bisects $\angle B$, $AB = 4$, $BC = 6$, and $AC = 8$. Find $DC$.

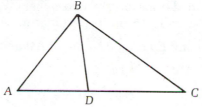

6. The sides of a triangle are 20, 30, and 35 cm. Find the two segments into which the bisector of the largest angle divides the opposite side.

7. In $\triangle ABC$, $AD$ bisects $\angle A$. If $BC = 12$, $BD = 4\frac{4}{5}$, and $AB = 6$, find $AC$.

8. In $\triangle ABC$, $BD$ bisects $\angle B$. If $AB = 4$, $BC = 8$, and $AC = 7$, find $DC$.

9. The bisector of $\angle A$ of $\triangle ABC$ divides $BC$ into segments whose ratio is 5 to 2 with the larger segment adjacent to $AB$. If $AB = 12$, find the length of $AC$.

10. If $AB = 5$, $AD = 3$, $AC = 8\frac{1}{3}$, $AB \perp BC$, $BD \perp AC$. Find $DC$, $BD$, and $BC$.

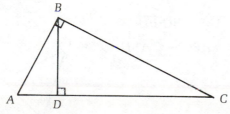

11. Using the figure for #10, if $BD = 12$, $AD = 5$, $CD = 28\frac{4}{5}$, find $AB$ and $BC$.

12. Using the figure for #10, if $BD = 8$, $DC = 15$, $BC = 17$, find $AD$, $AC$, $AB$.

**157**

13. If an 8 m billboard is divided by lines such that $AB = 200$ cm, $BC = 180$ cm, $CD = 300$ cm, and $DE = 120$ cm, find the length of each segment of a 10 m and a 12 m transverse line.

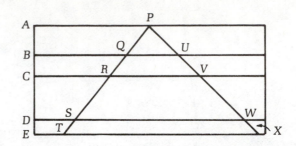

14. On a yacht, the fore and aft lines to the mast form a right angle half way up the mast. If the yacht is 25 m long and the mast is 24 m tall, how long is each line?

15. A method used by carpenters to divide a board into equal parts is to use the vertical studding of a building as parallel lines, and to place the board to be divided transversely across them. Why does this work? Note—studding is usually 16 in. from center to center.

16. Given: $D$ and $E$ are midpoints of $AC$ and $BC$ respectively

   Prove: $\triangle ABC \sim \triangle CDE$

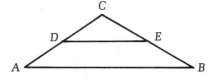

17. Given: $CD \perp AB$, $AD = 5$, $CD = 10$, $BD = 20$

   Prove: $\triangle ACD \sim \triangle BCD$

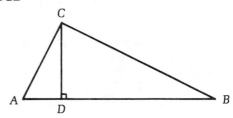

18. Given: $\overline{AB}^2 = \overline{BC} \cdot \overline{BD}$

   Prove: $\triangle ABC \sim \triangle ABD$

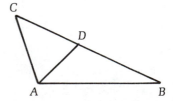

19. Prove that two right triangles are similar, if an acute angle of one is equal to an acute angle of the other.

20. **Given:** Rectangle *ABCD* (in a rectangle all angles are right angles and the opposite sides are parallel), *DE* ⊥ *AC*

    **Prove:** △*CDE* ~ △*ABC*

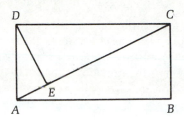

21. The sides of one triangle are 3, 6⅞, and 5.
    The sides of a second triangle are 7, 11⅔, and 16.
    Are the triangles similar?

22. ∡*A* = ∡*D* and ∡*B* = ∡*E*. Find *DE* and *EF*.

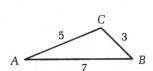

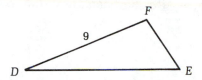

23. A man 180 cm tall has a shadow of 210 cm at the same time that a tree has a shadow of 32 m. How tall is the tree?

24. **Given:** ∡*P* is supplementary to ∡*STQ*

    **Prove:** $\dfrac{RT}{PR} = \dfrac{RS}{QR}$

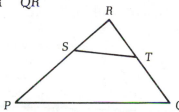

*25. **Given:** *CD* bisects ∡*ACB*

    $\dfrac{AC}{CD} = \dfrac{CE}{BC}$

    **Prove:** ∡*A* = ∡*E*

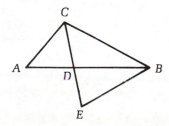

*26. Draw any acute angle. Put a point anywhere inside the angle. Now construct a line segment from one side of the angle to the other side that will be bisected by the point.

27. We wish to find the height of a flagpole on an overcast day when there are no shadows. Put a mirror on the ground and stand in such a position that you see the top of the flagpole in the mirror. If you know your own height, how will you find the height of the flagpole?

28. An aerial photograph, taken from a height of 5 km, shows a river known to be 90 m wide as 3.2 cm wide in the photograph. The following week another picture is taken (with the same camera) from an altitude of 4 km. The river now measures 2.8 cm in the photograph. How wide is the river now?

*29. A story is told in Polya's book "*Mathematical Methods in Science*" about the residents of a Greek city and their solution to a drought. The city (*C*) was on the other side of a mountain from a water supply (*S*). They decided to tunnel through the mountain to bring the water to the city. To save time crews were to begin digging from both sides and meet in the middle. From a distant point it was possible to see both the city and the water supply, and measure the distance to each. How did they determine the exact direction for each crew to begin digging?

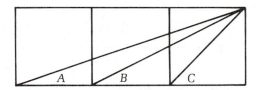

observation point

*30. *Given:* Three squares with diagonals

*Prove:* $\angle A + \angle B = \angle C$

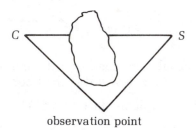

*31. A farmer wishes to measure the distance, *SW*, across a swamp on his property. He stands at some convenient point *M* on dry ground and locates some point *P*, in line with S on the other side of the swamp. He then constructs *PA* parallel to *SW*. When he measures he finds that $WM = 85\,\text{m}, AM = 20\,\text{m}, \text{and } PA = 17\,\text{m}$. How wide, to the nearest meter, is the swamp? Write a proof to justify your answer.

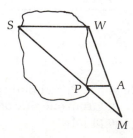

*32. *Prove:* Two triangles whose sides are respectively parallel are similar.

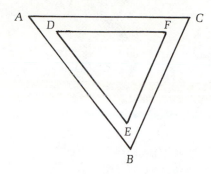

*Hint:* Use this figure, extend *DE* to *BC*, extend *DF* to *BC*, look for some corresponding angles.

*33. *Prove:* Two triangles whose sides are respectively parallel are similar.

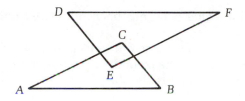

*Hint:* Use this figure, extend *AC* and *BC* to *DF*, and look for some corresponding angles.

## 16. The Pythagorean theorem

This theorem, and its corollaries, is probably the best known and most used of all the theorems of Euclidean geometry. It will be used frequently in trigonometry, calculus, and other future course work, as well as having many practical everyday applications.

A special case of this theorem was known to the Egyptians as long ago as 2000 BC. They realized that segments of 3, 4, and 5 units on a rope could be used to form a right angle. This relationship was used by the Egyptian "rope stretchers" for their surveying. As early as 1000 BC the Babylonians realized the general relationship that in a right triangle $a^2 + b^2 = c^2$, where $a$ and $b$ are the legs and $c$ is the hypotenuse. Of course, they had arrived at this conclusion inductively, and had not *proven* that it was true in all cases.

Pythagoras, about 500 BC, is believed to be responsible for the first deductive proof of this relationship. The proof given by Euclid in *The Elements*, about 300 BC, is probably that of Pythagoras. Many different

Line, graph, and hypotenuse.

proofs are possible. One book has been published with 256 different proofs.*

The original proof dealt with the concept of areas. The theorem was stated "In a right triangle the square *on* the hypotenuse is equal to the sum of the squares *on* the two legs".

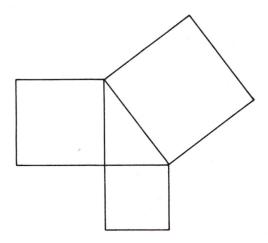

We will return to the Pythagorean Theorem as a statement about areas after we discuss area. Many clever and interesting proofs are possible using equivalent areas. Our present approach to the theorem will be with regard to numerical relationships. We refer to the legs and the hypotenuse as numerical measures and consider the squares *of* these numbers.

*The Pythagorean Proposition, Elisha Scott Loomis, National Council of Teachers of Mathematics, 1940.

> *In a right triangle the square of the hypotenuse is equal to the sum of the squares of the two legs.*

Given: Right triangle $ABC$

Prove: $a^2 + b^2 = c^2$

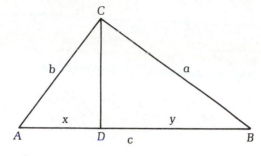

| | |
|---|---|
| **1.** Draw $CD \perp AB$ | **1.** Construction |
| **2.** $\triangle ABC \sim \triangle BCD$ | **2.** Alt. to hypotenuse of a rt. $\triangle$ makes $\sim \triangle$s |
| **3.** $\dfrac{c}{a} = \dfrac{a}{y}$ | **3.** Corr. parts of $\sim \triangle$s |
| **4.** $\triangle ABC \sim \triangle ACD$ | **4.** Alt. to hypotenuse of a rt. $\triangle$ makes $\sim \triangle$s |
| **5.** $\dfrac{c}{b} = \dfrac{b}{x}$ | **5.** Corr. parts of $\sim \triangle$s |
| **6.** $a^2 = cy$ | **6.** Prod. of means = prod. of ext. |
| **7.** $b^2 = cx$ | **7.** Prod. of means = prod. of ext. |
| **8.** $a^2 + b^2 = cx + cy$ | **8.** Addition |
| **9.** $a^2 + b^2 = c(x + y)$ | **9.** Algebraic substitution |
| **10.** $a^2 + b^2 = c^2$ | **10.** Substitution |

Here are some examples using this most important theorem:

**How long is the diagonal of a square with sides of 4 cm? A figure always helps:**

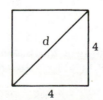

Since a square has right angles the theorem will apply and

$$d^2 = 4^2 + 4^2 \qquad d^2 = 2 \cdot 16 \qquad d = 4\sqrt{2} \text{ cm}$$

How long is the diagonal of a rectangle with sides of 6 cm and 8 cm? Draw a figure:

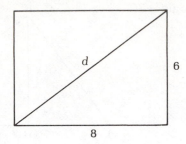

Again we have right angles so the Pythagorean Theorem applies:

$$d^2 = 6^2 + 8^2 \qquad d^2 = 36 + 64 \qquad d^2 = 100 \qquad d = 10 \text{ cm}$$

A 6 m ladder leans against a vertical wall with its bottom 2 m from the wall. How high on the wall will it reach?

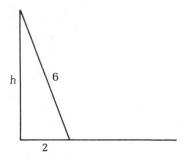

A vertical wall implies right angles so we can use the theorem, but this time it is a leg that is the unknown:

$$6^2 = h^2 + 2^2 \qquad 36 = h^2 + 4 \qquad h^2 = 32 \qquad h = 4\sqrt{2} \text{ m}$$

If an approximate answer would be more useful than an exact answer $h \approx$ 5.66 m (The number of digits you are entitled to depends on the accuracy of your original measurements.)

Will a 50 cm ruler lie flat in a drawer 25 cm × 40 cm? Assuming the drawer is rectangular, we need to find the maximum dimension—which is the diagonal.

$$d^2 = 25^2 + 40^2 \qquad d^2 = 625 + 1600 \qquad d^2 = 2225 \qquad d = \sqrt{2225}$$

which is less than 50 ($50 = \sqrt{2500}$), so it will not fit.

The equilateral triangle, the hexagon, and other forms of the 30–60–90 triangle are very common in construction, tiling, and trigonometric problems; so much so that the student should memorize the relationship given by the following corollary:

> *The ratio of the sides of a 30–60–90 triangle is $1 : \sqrt{3} : 2$.*

Given: 30–60–90 triangle $PQR$

Prove: $\dfrac{a}{1} = \dfrac{b}{\sqrt{3}} = \dfrac{c}{2}$

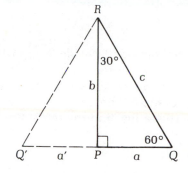

1. Reflect $PQR$ about $PR$
2. $\triangle PQR$ is 30–60–90
3. $\triangle Q'RQ$ is equiangular
4. $\triangle Q'RQ$ is equilateral
5. $a' + a = c$
6. $a' = a$
7. $2a = c$
8. $a = \dfrac{c}{2}$
9. $a^2 + b^2 = c^2$
10. $a^2 + b^2 = (2a)^2$
11. $b^2 = 3a^2$
12. $b = \sqrt{3}a$
13. $a = \dfrac{b}{\sqrt{3}}$
14. $\dfrac{a}{1} = \dfrac{b}{\sqrt{3}} = \dfrac{c}{2}$

<br>

1. Existence of a reflection
2. Given
3. Properties of reflection
4. Equiangular $\triangle$ is equilateral
5. Def. of equilateral
6. Property of reflection
7. Substitution
8. Division
9. Pythagorean Th.
10. Substitution
11. Subtraction
12. Roots of equals
13. Division
14. Substitution

Remember this theorem as a triangle with the following ratio of sides:

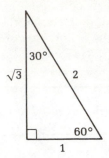

Note that the smallest side is opposite the smallest angle and the largest side is opposite the largest angle. Since $\sqrt{3}$ is approximately 1.732 (which is a good number to remember) it is the middle sized side and is opposite the 60° angle.

Using the ratios, find the remaining two sides when one side is given:

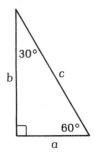

$a = 7$

$$\frac{7}{1} = \frac{b}{\sqrt{3}} \quad b = 7\sqrt{3} \quad ; \quad \frac{7}{1} = \frac{c}{2} \quad c = 14$$

$c = 10$

$$\frac{a}{1} = \frac{10}{2} \quad a = 5 \quad ; \quad \frac{5}{1} = \frac{b}{\sqrt{3}} \quad b = 5\sqrt{3}$$

$b = 9\sqrt{3}$

$$\frac{a}{1} = \frac{9\sqrt{3}}{\sqrt{3}} \quad a = 9 \quad ; \quad \frac{9}{1} = \frac{c}{2} \quad c = 18$$

$b = 7$

$$\frac{a}{1} = \frac{7}{\sqrt{3}} \quad a = \frac{7}{\sqrt{3}} = \frac{7\sqrt{3}}{3} \quad ; \quad \frac{\frac{7\sqrt{3}}{3}}{1} = \frac{c}{2} \quad c = \frac{14\sqrt{3}}{3}$$

   Another much used ratio is the ratio of the sides of an isosceles right triangle. This has applications as the diagonal of a square, the most efficient support for a vertical, the distance from home plate to second base, and many, many more applications.

*The ratio of the sides of an isosceles right triangle is $1:1:\sqrt{2}$.*

Given:  Isosceles right triangle $ABC$

Prove:  $\dfrac{a}{1} = \dfrac{b}{1} = \dfrac{c}{\sqrt{2}}$

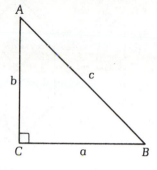

1. $\triangle ABC$ is isosceles rt. triangle    1. Given

2. $a = b$    2. Def. of isosceles

3. $a^2 + b^2 = c^2$    3. Pythagorean Th.

4. $a^2 + a^2 = c^2$    4. Substitution

5. $2a^2 = c^2$    5. Substitution

6. $c = \sqrt{2}\,a$    6. Roots of equals

7. $a = \dfrac{c}{\sqrt{2}}$    7. Division

8. $\dfrac{a}{1} = \dfrac{b}{1} = \dfrac{c}{\sqrt{2}}$    8. Substitution

Using the ratios, find the remaining two sides when one side is given:

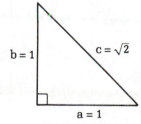

$a = 9$

$\dfrac{9}{1} = \dfrac{b}{1}$    $b = 9$  ; $\dfrac{9}{1} = \dfrac{c}{\sqrt{2}}$    $c = 9\sqrt{2}$

$c = 11\sqrt{2}$

$a = b = \dfrac{11\sqrt{2}}{\sqrt{2}}$    $a = b = 11$

$c = 13$

$a = b = \dfrac{13}{\sqrt{2}} = \dfrac{13\sqrt{2}}{2}$

This tiling pattern, a very common one, is composed of isosceles right triangles. Can you find different sized triangles and squares that illustrate the Pythagorean Theorem?

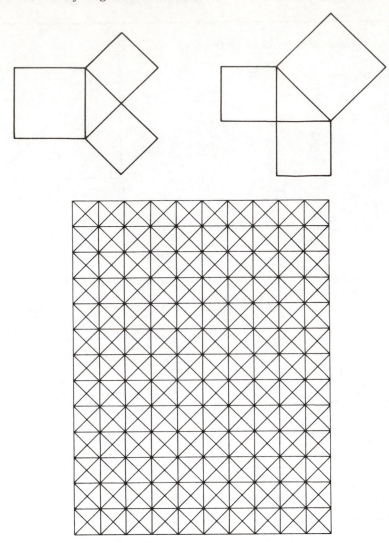

It is interesting to note that the Pythagorean Theorem can now be used to construct the length of the square root of any number (which could then be measured to obtain a numerical approximation). One method of doing this is a recursion technique: that is, each result is dependent upon a previous result. Begin with an isosceles right triangle. Its hypotenuse will be $\sqrt{2}$. Now, using the hypotenuse as a leg, construct another leg of 1 and draw the new hypotenuse. It will have a measure of $\sqrt{3}$.

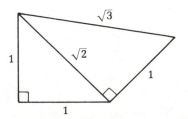

Now, using this new hypotenuse as a leg, construct another leg of 1 and draw the new hypotenuse. It will measure $\sqrt{4}$, or 2.

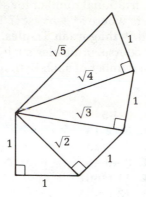

Continuation of this process will give hypotenuses of $\sqrt{5}$, $\sqrt{6}$, and so on. Of course, if you wished to construct the $\sqrt{137}$ this method would be tedious. What we need is an explicit method, a method that does not depend upon a previous result. This can be done, but requires a little more creativity.

We can construct the square root of a particular number, $X$, if we can find two squares (1, 4, 9, 16, 25, . . .) which have either a sum or a difference of $X$. If the sum of the two numbers is $X$, use the square roots of these numbers as legs of a right triangle and the hypotenuse will be $\sqrt{X}$. If the difference of the two squares is $X$, use the square root of the smaller one as a leg and the square root of the larger one as the hypotenuse and the remaining leg will be $\sqrt{X}$. To illustrate—if we wished to construct $\sqrt{17}$, since $16 + 1 = 17$, construct a right triangle with legs of 1 and 4 and the hypotenuse is $\sqrt{17}$. If we wish to construct $\sqrt{137}$, since $121 + 16 = 137$, construct a right triangle with legs of 11 and 4 and the hypotenuse will be $\sqrt{137}$.

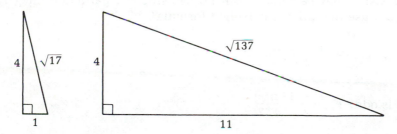

To construct $\sqrt{15}$, since $16 - 1 = 15$, construct a right triangle with a leg of 1 and hypotenuse of 4, and the other leg will be $\sqrt{15}$. To construct $\sqrt{57}$, since $29^2 - 28^2 = 57$, construct a right triangle with a leg of 28 and a hypotenuse of 29, and the other leg will be $\sqrt{57}$.

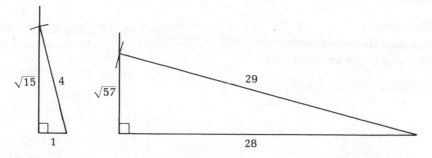

The Pythagorean right triangle of the Egyptian "rope stretchers" had three integers as its sides, 3-4-5, whereas most of the triangles we have been using have an irrational number for at least one side. Are there other sets of three *integers* where $a^2 + b^2 = c^2$? The answer is "yes," and such numbers are called **Pythagorean Triples.** As examples—5-12-13, 7-24-25, 9-40-41 and, of course, multiples such as 6-8-10 or 10-24-26. In fact, there are an infinite number of such Pythagorean Triples. Consider this list:

**1.** 3-4-5  $\quad\quad = X_1\text{-}Y_1\text{-}Z_1$
**2.** 5-12-13
**3.** 7-24-25
**4.** 9-40-41
**5.** 11-60-61
**6.** 13-84-85
**7.** 15-112-113

This list can be continued by applying the recursion formula: For the nth triplet $(X_n\text{-}Y_n\text{-}Z_n)$ we use:

$$X_n = X_{n-1} + 2 \quad\quad Y_n = Y_{n-1} + 4n \quad\quad Z_n = Y_n + 1.$$

So for the eighth Pythagorean Triple:

$$X_8 = X_7 + 2 = 15 + 2 = 17$$
$$Y_8 = Y_7 + 4(8) = 112 + 32 = 144$$
$$Z_8 = Y_8 + 1 = 144 + 1 = 145$$

**8.** 17-144-145

Since this recursion formula can be applied over and over, the list is infinite. We have an infinite number of triples.

Yet this formula only gives you *some* of the Pythagorean Triples. If you wish to have a Pythagorean Triple using any particular integer, n, we can use the following explicit formula:

if n is even: $n, \dfrac{n^2 - 4}{4}, \dfrac{n^2 + 4}{4}$

if n is odd:  $n, \dfrac{n^2 - 1}{2}, \dfrac{n^2 + 1}{2}$

**To illustrate: if** $n = 20, \dfrac{n^2 - 4}{4} = \dfrac{400 - 4}{4} = 99, \dfrac{n^2 + 4}{4} = 101$

**so the Pythagorean Triple is 20-99-101.**

**If** $n = 11, \dfrac{n^2 - 1}{2} = \dfrac{121 - 1}{2} = 60, \dfrac{n^2 + 1}{2} = \dfrac{121 + 1}{2} = 61,$ **so the Pythagorean Triple is 11-60-61.**

This formula does not give all of the Pythagorean Triples either! For instance, there is another triple using 20—20-21-29—which is not the one which the formula gives.

To find *all* Pythagorean Triples with a given leg, $a$: list all of the integral factor pairs of $a^2$, call them $(f_1, f_2)$. Eliminate those whose sum is odd. The other leg is then $\dfrac{f_2 - f_1}{2}$, and the hypotenuse is $\dfrac{f_2 + f_1}{2}$.

**Example: Find all Pythagorean Triples with a leg of 10.** $10^2 = 100$, so factors of 100 are: (1, 100), (2, 50), (4, 25), and (5, 20). Eliminating those with odd sums leaves only (2, 50). Now $\dfrac{50 - 2}{2} = 24$ and $\dfrac{50 + 2}{2} = 26$. Therefore, (10-24-26) is the only Pythagorean Triple with a leg of 10.

**Example: Find all Pythagorean Triples with a leg of 24.** $24^2 = 576$, so factors of 576 are: (1, 576), (2, 288), (3, 192), (4, 144), (6, 96), (8, 72), (9, 64), (12, 48), (16, 36), (18, 32). Eliminating the (1, 576), (3, 192), and (9, 64) because their sums are odd leaves us with:

$$\frac{288 - 2}{2} = 143, \quad \frac{288 + 2}{2} = 145 \qquad \frac{144 - 4}{2} = 70, \quad \frac{144 + 4}{2} = 74$$

$$\frac{96 - 6}{2} = 45, \quad \frac{96 + 6}{2} = 51 \qquad \frac{72 - 8}{2} = 32, \quad \frac{72 + 8}{2} = 40$$

$$\frac{48 - 12}{2} = 18, \quad \frac{48 + 12}{2} = 30 \qquad \frac{36 - 16}{2} = 10, \quad \frac{36 + 16}{2} = 26$$

$$\frac{32 - 18}{2} = 7, \quad \frac{32 + 18}{2} = 25$$

**Therefore, the Pythagorean Triples with a leg of 24 are: (24-143-145), (24-70-74), (24-45-51), (24-32-40), (24-18-30), (24-10-26), (24-7-25).**

There are 50 Pythagorean Triples in which all three sides are less than 100. This list includes multiples such as 6-8-10 and 10-24-26. The ones not given by either of the first two formulas discussed are:

| | | | | |
|---|---|---|---|---|
| 20-21-29 | 28-45-53 | 33-56-65 | 36-77-85 | 39-80-89 |
| 48-55-73 | 65-72-97 | | | |

Can you complete the list of 50?

The following is a very clever little puzzle that is based on the Pythagorean Theorem. It uses the area concept of the squares on the legs being equal to the square on the hypotenuse. The problem is to take the 5 pieces which form the squares on the two legs and rearrange them to form a square on the hypotenuse. It will make a more usable puzzle if you copy the pieces on heavy paper or wood and cut them out. Not having seen the puzzle as given, some people will have difficulty forming the square on the longer leg of the right triangle, and even more difficulty forming the square on the hypotenuse.

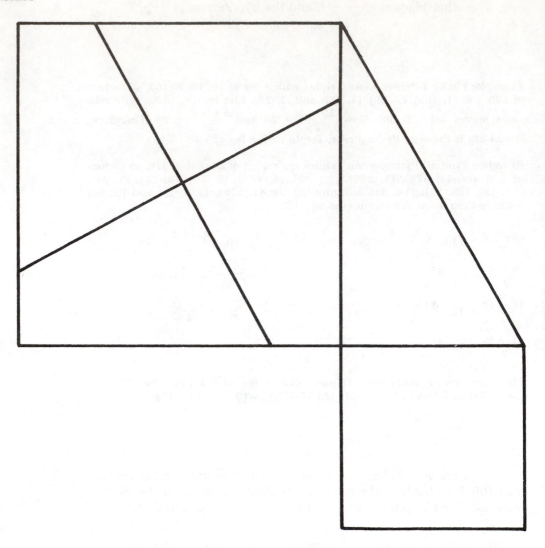

As a final comment, the theorem is sometimes stated so as to refer to sides rather than legs: "The square on the hypotenuse is equal to the sum of the squares on the other two sides."

If Pythagoras had been a Native American rather than an Ancient Greek, his discovery of the theorem might have come about in this manner:

Once upon a time, in an Indian village, three women who were in the family way all went to the maternity tepee at the same time, to await the blessed event. The medicine man made them comfortable by putting them on soft beds of hides. The first woman was on a bed of elk hides, the second woman on a bed of deerskins, and the third woman on a hippopotamus skin that the medicine man had found somewhere. After a while, the medicine man came out of the tepee to announce to the anxious braves that the first woman had a 6 lb. boy. A little later, he returned with the announcement that the second woman had given birth to an 8 lb. son. And still later, he returned a third time with the happy announcement that the third woman had given birth to twin 7 lb. sons. Which proves that "the squaw on the hippopotamus is equal to the sons of the squaws on the other two hides!"—Ugh!

1. The hypotenuse of a right triangle is 9 cm and one of the legs is 5 cm. Find the length of the other leg.

2. The hypotenuse of a right triangle is 20 cm and one of the legs is 16 cm. Find the length of the other leg.

3. The legs of a right triangle are 8 cm and 12 cm. Find the hypotenuse.

4. Find the diagonal of a square whose side is 5 cm.

5. Find the diagonal of a rectangle whose dimensions are 4 cm by 6 cm.

6. Find the side of an equilateral triangle whose altitude is 12 cm.

7. Find the legs of an isosceles right triangle whose hypotenuse is 14 cm.

8. The altitude upon the hypotenuse of a right triangle divides the hypotenuse into segments of 2 cm and 30 cm. Find the shorter leg of the triangle.

9. Find the perimeter of a 30–60–90 triangle whose longer leg is 6 cm.

10. The legs of an isosceles triangle are each 34 cm and the base is 60 cm. Find the altitude upon the base.

11. The legs of a right triangle are 15 cm and 20 cm. The altitude is drawn to the hypotenuse. Find the segments of the hypotenuse.

12. In the problem above, find the altitude.

13. A tree is broken 24 m above the level ground and the top rests on the ground at a distance of 18 m from the bottom of the tree. How tall was the tree before it broke?

14. A man starting at point $A$ walks 10 km east to point $B$, then 8 km north to point $C$, then 5 km east to point $D$. How far is he from his starting point?

15. Find the side of a square whose diagonal is 18 cm.

16. A ladder is both difficult to use and unsafe if its foot is placed too close to a wall. Assume that a 5 m ladder is placed against a wall so that it makes an angle of 60° with the ground. How high will it reach on the wall?

17. If a 5 m ladder is placed against a wall so that its foot is 2 m from the wall, how high up the wall will it reach?

18. A 10 m flagpole is to be supported by three guy-wires attached at the 7 m level and making a 45° angle with the ground. How much wire is necessary?

19. If the wires above are anchored so as to make a 60° angle with the ground, how much wire is necessary?

20. When buying framing material for the construction of picture frames, a good rule-of-thumb is to buy a length equal to the perimeter of the picture plus eight times the width of the frame (plus a small amount

for the width of the saw cuts). Draw a figure and show why this rule works.

21. We wish to frame a 60 cm by 48 cm picture with material that is 8 cm wide. How much framing material should we buy? Add 2 cm for the width of saw cuts. (See problem 20.)

22. List all Pythagorean Triples with a leg of 16.

23. Find a Pythagorean Triple with a leg of 29.

24. In laying out a foundation for an 8′ by 10′ storage shed, the builder wants to be certain the foundation is a rectangle and not a parallelogram.

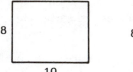

He can do this by checking the length of a diagonal. How long should they be?

25. Four holes are to be drilled in a circular plate with a diameter of 15 cm. The holes are to be drilled 3 cm from the edge of the plate. What is the center to center distance for two adjacent holes?

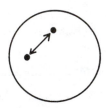

26. What is the minimum diameter tree trunk that can be used to mill a square beam 45 cm on a side?

27. Which of the following sets of three integers are Pythagorean Triples?

| | |
|---|---|
| 1-2-3 | 8-15-17 |
| 2-3-4 | 8-10-12 |
| 3-4-5 | 12-16-20 |
| 6-8-10 | 15-20-25 |
| 7-9-12 | 20-21-29 |
| 7-24-25 | 36-77-85 |

28. Find a right triangle (there are many) that has a leg of 7 and the hypotenuse and other leg are integers. Find a right triangle with a leg of 11; a leg of 13; a leg of 17; a leg of 23.

29. Use the Pythagorean Theorem to find the distance between the points (2, 3) and (5, 7).

30. Use the Pythagorean Theorem to find the distance between the points (−5, −3) and (2, 8).

31. Use the Pythagorean Theorem to find the distance between the points $(X_1, Y_1)$ and $(X_2, Y_2)$.

*32. Find the longest interior dimension (called the space diagonal) of a box measuring 1 m by 1.5 m by 2 m.

*33. Find the general formula for the space diagonal of a rectangular solid with edge dimensions of $a$, $b$, and $c$.

*34. Two cities, Ourtown and Theirtown, lie on opposite sides of a river. The river is straight and 200 m wide. Ourtown is 2 km from the river and Theirtown is 4 km from the river. Theirtown is 6 km up the river from Ourtown. The city councils of the two towns have decided to share equally the cost of joining the two cities with a road and a bridge. The bridge is to be built perpendicular across the river and roads joining it to each of the towns are to be the same length. How far from each town will the bridge be?

*35. A 5 in. diameter air conditioning pipe is to be held in a corner by a 1 in. board. What is the shortest length of board that will do the job?

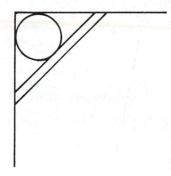

*36. A boy bug is located on the wall of a room at B. A girl bug is on an adjacent wall at G. What is the shortest distance from B to G?

B is 1 m from the ceiling and 1 m from the wall.
G is 2 m from the ceiling and 1 m from the wall.

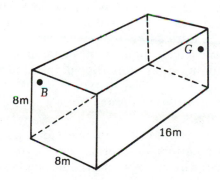

**4–Summary**

Theorems on angle relationships:

The sum of the angles of a triangle is 180°.
An exterior angle of a triangle is equal to the sum of the opposite interior angles.
An exterior angle of a triangle is greater than either opposite interior angle.
If two angles of one triangle are equal to two angles of another triangle, then the third angles are equal.
The acute angles of a right triangle are complementary.
Each angle of an equilateral triangle is 60°.

Ratio: the comparison of two quantities by division: $\dfrac{a}{b}$

Proportion: the equality of two ratios

Mean proportional: when the same number is used for both the second and third terms of the proportion—the $b$ in $\dfrac{a}{b} = \dfrac{b}{c}$

Similarity: a transformation which preserves ratios of distances.

Dilatation: a transformation that transforms each line into a parallel line.

Half-turn: a dilatation in which the center of similitude is between the two figures.

Theorems on similarity and proportion:

A line through two sides of a triangle parallel to the third side cuts off a similar triangle.
A line through two sides of a triangle parallel to a third side divides the sides proportionally.
Three or more parallel lines cut off proportional segments on any two transversals.
If two angles of one triangle are equal to two angles of another triangle, then the triangles are similar. (AA, sim.)
If a line divides two sides of a triangle proportionally, then it is parallel to the third side.
If two triangles have an angle of one equal to an angle of the other and the including sides proportional, then the triangles are similar. (SAS, sim.)
If two triangles have their sides respectively proportional, then they are similar. (SSS, sim.)
The bisector of an angle of a triangle divides the opposite side into segments proportional to their adjacent sides.
The altitude to the hypotenuse of a right triangle forms three similar triangles.
The altitude to the hypotenuse of a right triangle is the mean proportional to the segments of the hypotenuse.

The Pythagorean Theorem and corollaries:

In a right triangle, the square of the hypotenuse is equal to the sum of the squares of the two legs.

The ratio of the sides of a 30–60–90 triangle is $1 : \sqrt{3} : 2$.

The ratio of the sides of an isosceles right triangle is $1 : 1 : \sqrt{2}$.

The methods of construction of irrational number lengths:

Use Pythagorean Triples for the sides of a right triangle.

Formulas for finding Pythagorean Triples $(X_n, Y_n, Z_n)$:

Starting with a known triple let
$$X_n = X_{n-1} + 2 \qquad Y_n = Y_{n-1} + 4n \qquad Z_n = Y_n + 1$$
Or, if $n$ is even:

$$\left( n, \ \frac{n^2 - 4}{4}, \frac{n^2 + 4}{4} \right)$$

if $n$ is odd:

$$\left( n, \ \frac{n^2 - 1}{2}, \frac{n^2 + 1}{2} \right)$$

The corporate offices of Fluor Corporation in Irvine, California, are a hexagonal prism. *Courtesy Fluor Corporation.*

# CHAPTER 5

# POLYGONS AND AREA

In this chapter, we will classify the polygons as to their number of sides and develop some theorems about the measure of their angles. We will look at the problem of which polygons are constructible and which are not. We will see how to construct some of the ones for which a construction is possible.

Of special interest in this chapter are the many different types of four-sided polygons—the quadrilaterals. We will study the unique properties of each one, and develop some theorems about them. We will then exercise our ingenuity with problems of constructing specific quadrilaterals when only a minimum of information is given. Finally, we will develop formulas for the areas of various quadrilaterals.

The material on concurrency can be covered here, if time permits.

Polygon.

## 17. Polygons and polygon constructions

Although some naive students might think a polygon was a dead parrot, a proper geometrical definition would more likely be that a **polygon** is a closed plane figure formed by three or more line segments. This broad definition includes many different kinds of figures:

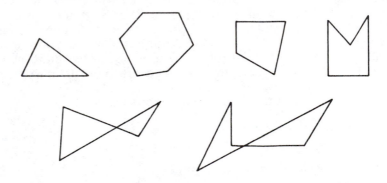

In order to discuss, classify, and learn something about such a variety of figures we need to develop some vocabulary. Although many of the relationships we will derive can be extended to polygons where the line segments cross each other, we will restrict our discussion to non-crossing polygons. Of these, there are two general types, concave and convex. In a **concave** polygon at least one of the interior angles is greater than 180°.

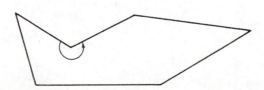

A **convex** polygon is one in which all of the interior angles are less than 180°. In the discussions which follow, we will work only with convex polygons.

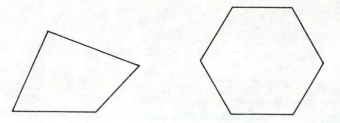

An **equilateral** polygon is one in which all of the sides are equal. An **equiangular** polygon is one in which all of the angles are equal. Notice that, unlike in a triangle, one does not imply the other. A polygon can be equilateral and not equiangular:

Or, a polygon can be equiangular and not equilateral:

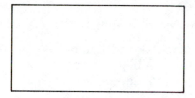

A **regular** polygon is one which is both equilateral and equiangular.

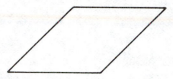

Some of the terms used with a polygon are:

a **side**—one of the line segments forming the polygon, $AB$;
an **interior angle**—formed by the intersection of two sides, $\angle ABC$;
a **diagonal**—a line segment joining two non-adjacent vertices, $AC$;
the **perimeter**—the sum of all the sides, $ACBDEA$; and
an **exterior angle**—formed by one of the sides and an adjacent side produced, $\angle EAF$.

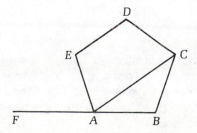

The Pentagon. *Courtesy United States Department of the Navy.*

Polygons are named according to their number of sides:

| | | |
|---|---|---|
| 3—triangle | 4—quadrilateral | 5—pentagon |
| 6—hexagon | 7—heptagon | 8—octagon |
| 9—nonagon | 10—decagon | 12—dodecagon |

Would it be proper to call a polygon with an infinite number of sides a circle?

Before we prove some theorems about angle measures in polygons, it might be instructive to first do some experimentation: We know that the sum of the angles of a triangle is 180°. We also know that the sum of the angles of a square or a rectangle is 4 times 90° or 360°. But is the sum of the angles of *any* quadrilateral 360°? What about one like this:

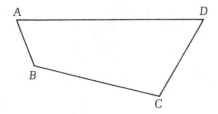

Will measurement with a protractor decide the question? It will not, since all measurements are only approximations. But if you draw diagonal *AC* you will then have two triangles, the sum of whose angles is 2 times 180°, or 360°. Further investigation of other polygons might give the following information:

| # of Sides | Sum of the Angles |
|:---:|:---:|
| 3 | 180 |
| 4 | 360 |
| 5 | 540 |
| 6 | 720 |
| 7 | 900 |
| 8 | 1080 |
| n | 180(n − 2) |

The following theorem gives a general formula for this relationship between the number of sides and the sum of the angles.

---

> *The sum of the interior angles of a polygon of n sides is 180(n − 2).*

Given: ABCDE . . . is a polygon of n sides

Prove: $\angle ABC + \angle BCD + \angle CDE + \cdots = 180(n - 2)$

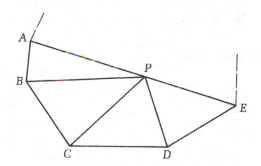

| | |
|---|---|
| **1.** From any point P in the interior of the polygon draw PA, PB, PC, . . . | **1.** Construction |
| **2.** ABCDE . . . is a polygon of n sides | **2.** Given |
| **3.** Polygon is divided into n △s | **3.** There is a triangle on each side |
| **4.** The sum of the ∢s of the △s = 180n | **4.** Multiplication |
| **5.** The sum of the angles at P = 360 | **5.** 360° about a point |
| **6.** $\angle ABC + \angle BCD + \angle CDE + \cdots$ = 180n − 360 | **6.** Subtraction |
| **7.** $\angle ABC + \angle BCD + \angle CDE + \cdots$ = 180(n − 2) | **7.** Substitution |

This theorem applies to any kind of polygon with any number of sides. If the polygon happens to be regular, then each angle is the same size and we have the following corollary:

> *In a regular polygon of n sides each interior angle is* $\dfrac{180(n-2)}{n}$.

Given: $ABCDE \ldots$ is a regular polygon of $n$ sides

Prove: $\angle A = \dfrac{180(n-2)}{n}$

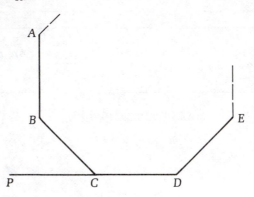

| 1. $ABCDE \ldots$ is a reg. polygon of $n$ sides | 1. Given |
|---|---|
| 2. $\angle A = \angle B = \angle C = \angle D = \cdots$ | 2. Def. of regular |
| 3. The sum of the interior angles $= 180(n-2)$ | 3. Previous Th. |
| 4. $n \angle A = 180(n-2)$ | 4. Substitution |
| 5. $\angle A = \dfrac{180(n-2)}{n}$ | 5. Division |

> *An exterior angle of a regular polygon of n sides is* $\dfrac{360}{n}$.

Given: $ABCDE \ldots$ is a regular polygon of n sides.

Prove: $\angle BCP = \dfrac{360}{n}$.  Use figure on page 184.

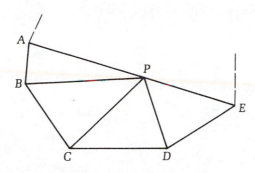

1. $ABCDE \ldots$ is reg. polygon of n sides

  1. Given

2. $\angle C = \dfrac{180(n-2)}{n}$

  2. Previous Th.

3. $\angle C + \angle BCP = 180$

  3. Form a st. angle

4. $\angle BCP = 180 - \angle C$

  4. Subtraction

5. $\angle BCP = 180 - \dfrac{180(n-2)}{n}$

  5. Substitution

6. $\angle BCP = \dfrac{360}{n}$

  6. Algebraic substitution

And one more final corollary:

> ### The sum of the exterior angles of a polygon is 360. (One exterior angle at each vertex.)

(Note that the number of sides is not mentioned—this is for all cases).

Given: Any polygon

Prove: Sum of ext. angles = 360

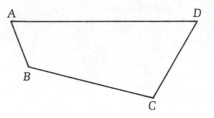

1. Each interior angle and its adjacent exterior angle are supplementary

1. Def. of ext. angle

2. A poly. of n sides will have n st. angles for a total of 180 times n degrees

2. Multiplication

3. The sum of the interior angles = 180(n − 2)

3. Previous Th.

4. The sum of the exterior angles = 180n − 180(n − 2)

4. Subtraction of (3) from (2)

5. The sum of the exterior angles = 360

5. Substitution (algebraically an equivalent expression)

To illustrate the use of the theorem and corollaries:

**Find the total number of degrees of the interior angles of a regular octagon.**

$$180(n − 2) = 180(8 − 2) = 1080$$

**Find the number of sides of a polygon in which the sum of the interior angles is 1800.**

$$180(n − 2) = 1800, \qquad n − 2 = 10, \qquad n = 12$$

**Find the number of degrees in one interior angle of a regular pentagon.**

$$\frac{180(n − 2)}{n} = \frac{180(5 − 2)}{5} = 108$$

**Find the number of sides of a regular polygon in which an interior angle is 120.**

$$\frac{180(n − 2)}{n} = 120, \qquad 180n − 360 = 120n, \qquad 60n = 360, \qquad n = 6$$

or: if the angle = 120, the ext. angle = 60 and $\frac{360}{n} = 60$, $\qquad n = 6$

Find the number of degrees in the exterior angle of a regular heptagon.

$$\frac{360}{n} = \frac{360}{7} = 51\frac{3}{7}$$

Find the number of sides of a regular polygon in which an exterior angle is 24.

$$\frac{360}{n} = 24, \qquad 24n = 360, \qquad n = 15$$

Find the sum of the exterior angles of a polygon of 193 sides.

360—sum of exterior angles for any polygon

Find the number of degrees in the fifth angle of a pentagon if the first four measure 119, 103, 83, 96.

$$180(n - 2) = 180(5 - 2) = 540$$
$$540 - 119 - 103 - 83 - 96 = 139$$

FRANK AND ERNEST                                    by Bob Thaves

Reprinted by permission of NEA.

Some regular polygons can be constructed with straight-edge and compass, and some cannot. It was not until March 29, 1796, that the problem was resolved by Carl Gauss (1777–1855). We will look at his theorem shortly. For 2200 years, since the time of the ancient Greeks, mathematicians had known how to construct regular polygons with 3, 4, 5, 6, 8, 10, and 15 sides. But they had tried and failed to construct polygons with 7, 9, 11, and 13 sides. Gauss was not yet 20 years old and was undecided whether to follow philology (the study of written records) or mathematics as his life's work. His resolution of this problem decided the question for him and he became one of the best and most prolific mathematicians of modern times. Gauss proved that a regular polygon is constructible when, and only when, the number of sides is $2^m(2^n + 1)$, where $m$ and $n$ are integers $\geq 0$ and $(2^n + 1)$ is a prime number or where the number of sides is the product of two numbers fitting this formula. For Example, if $m = 1$ and $n = 1$, $2^1(2^1 + 1) = 6$. Therefore, we can construct a regular hexagon. If $m = 0$ and $n = 4$, $2^0(2^4 + 1) = 1(17) = 17$. Gauss was the first to find a construction for the regular 17-sided polygon.* Notice that the formula requires $(2^n + 1)$ to be a prime number. It will be prime, if $n$ is a power of 2. If $n = 2^0, 2^1, 2^2, 2^3, 2^4 \ldots$; $(2^n + 1) = 2, 5, 17, 257, 65,537 \ldots$. Obviously, the problem of polygon construction is still unsolved, as no one has taken the time to find a way to construct a regular polygon of 65,537 sides—even though it has been proven possible.

**POLYGON
CONSTRUCTIONS**

---

*For Gauss' construction of the regular 17-gon see *Scientific American*, July, 1977; p. 122.

It is, therefore, possible to construct polygons of sides

3-6-12-24-etc., 4-8-16-32-etc., 5-10-20-40-etc., 15-30-60-etc.

The following construction methods are presented without proofs.
To construct a regular triangle:
  **Draw a circle with any radius.**
  **Using that same radius, mark off arcs around the circumference.**
  **Join every other mark.**

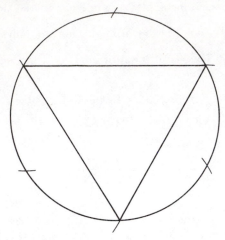

To construct a regular hexagon:

  **Join every mark.**

To construct a regular quadrilateral:

  **Draw a pair of perpendicular lines.**
  **Using their intersection as center, draw a circle.**
  **Join the points of intersection.**

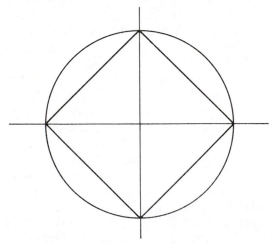

To construct a regular decagon:

  **Draw a circle with radius $AB$.**
  **Bisect $AB$, call it $E$.**
  **Construct a perpendicular at $B$.**
  **Mark $BE = BD$ and draw $AD$.**
  **Mark $DF = BE$.**
  **Mark $AC = AF$.**

(What we have done here is to construct a mean proportional, $\frac{AB}{AC} = \frac{AC}{BC}$. We will find other ways of doing this, and study its significance when we study the golden mean.)

   With compass set at $AC$, mark equal arcs on the circumference of the circle. Join these marks.

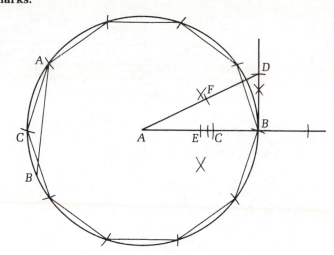

This construction must be done *very* carefully, as every small error is multiplied by a factor of 10 in the final step.

To construct a regular pentagon:

   Join every other mark in the decagon construction.

To construct a regular pentadecagon (15 sides) refer to the decagon construction:

   Lay off on the circle the radius $AB$ (which is also the side of a regular hexagon).
   From the same point lay off $AC$ (the side of the decagon).
   The difference, $BC$, will be the side of the pentadecagon.
   Mark it off on the circumference of the circle.
   Join the marks.

Notice how "round" a polygon with as few as 15 sides appears.

**A *false* construction of a regular heptagon** (With trigonometry we can prove that it is not exact—but it is accurate enough for most practical purposes.)

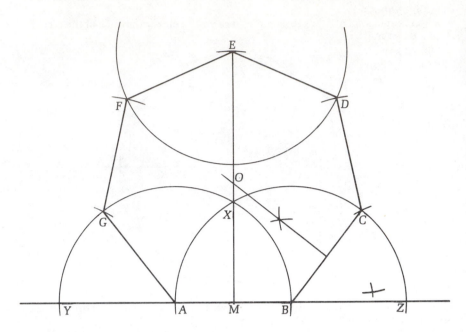

1. Draw circles *A* and *B* using radius *AB*. Circles intersect at point *X*. Circles intersect *AB* extended at *Y* and *Z*.
2. Draw perpendicular bisector *XM* of *AB*.
3. With center *Y* and radius *XM* draw the arc intersecting circle *A* at *G*. Draw *AG*.
4. With center *Z* and radius *XM* draw the arc intersecting circle *B* at *C*. Draw *BC*.
5. Construct the perpendicular bisector of *BC*, intersecting *XM* extended at *O*, the center of the heptagon.
6. With *O* as center and radius *OB*, locate *E* on *XM* extended.
7. With *E* as center and *AB* as radius, draw circle *E*.
8. With *G* as center and radius *AB*, draw arc to cross circle *E* at *F*; with *C* as center and radius *AB*, draw arc to cross circle *E* at *D*.
9. Complete regular heptagon *ABCDEFG*.

Snowflakes, despite individual variations, always form ice crystals in a regular hexagonal array. *Museum of Natural History.*

To construct any polygon with twice the number of sides of a given polygon:

> From the center of the circle draw radii to the ends of any side.
> Bisect the angle thus formed.
> The bisector will intersect the circle at the proper point for the side of the new polygon.
> Mark it off around the circumference.
> Join the marks.

We will illustrate by constructing a regular octagon from a regular quadrilateral:

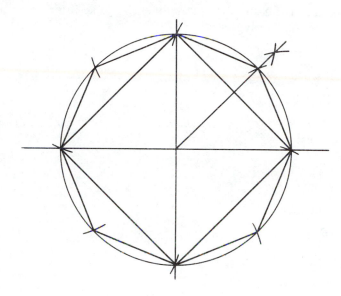

1–3. Find the sum of the interior angles of a polygon of:

1. 6 sides

2. 8 sides

3. 20 sides

4–6. Find the number of sides of a polygon if the sum of the interior angles is:

4. 1800°

5. 1260°

6. 540°

7–9. Find the number of degrees in each interior angle of a regular polygon of:

7. 5 sides

8. 9 sides

9. 12 sides

10–12. Find the number of sides of a polygon if each interior angle is:

10. 108°

11. 120°

12. 144°

13–15. Find the sum of the exterior angles of:

13. a pentagon

14. a polygon of 21 sides

15. a polygon of 100 sides

16–18. Find the number of degrees in each exterior angle of an equiangular polygon of:

16. 4 sides

17. 8 sides

18. 15 sides

19–21. Find the number of sides of a polygon if each exterior angle is:

19. 24°

20. 40°

21. 72°

22. If three angles of a quadrilateral are 75°, 85°, and 100°, find the fourth angle.

23–27. Which of the following can be an interior angle of a regular polygon?

23. 174°

24. 169°

25. 145°

26. 135°

27. 80°

28. Three exterior angles of a pentagon are 65°, 85°, and 90°, and the other two exterior angles are in the ratio of 1:2. Find each of the *interior* angles of the pentagon.

29. How many sides has a polygon if the sum of the interior angles is five times the sum of the exterior angles?

30. Find the number of sides of a regular polygon each of whose interior angles is 162°.

31. Find the number of degrees in each interior angle of a regular decagon.

32. Construct a regular dodecagon.

*33. **Prove:** Two triangles with their sides respectively perpendicular are similar. (Use this figure.)

> **Hint:** The sum of the angles of a quadrilateral is 360°; find some supplementary angles.

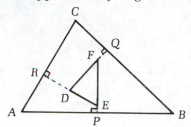

*34. Three straight lines can form, at most, one triangle. Four straight lines can form, at most, two non-overlapping triangles. Five straight lines can be arranged so as to form a maximum of five non-overlapping triangles. Show how this can be done.

*35. Show how to arrange six straight lines so as to form seven non-overlapping triangles. This problem is trivial if you have solved the previous problem, but nearly impossible if you have not.

**36. The number of non-overlapping triangles that can be formed by n lines is unknown. Perhaps you would like to derive the formula and become famous!

*37. If the corners of a square of side 12 cm are cut off to form a regular octagon, what will be the length of the side of the octagon?

*38. Find the sum of $\angle A + \angle B + \angle C + \angle D + \angle E + \angle F$. PQRSTU is any convex hexagon.

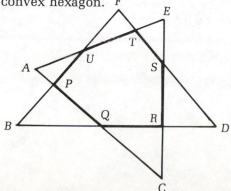

*39.  Games like tic-tac-toe lose their interest when both parties play perfectly and every game is a draw. Games like Nim lose their interest when the winning strategy is known by one or both of the players. Here is a very simple game which never ends in a draw, and for which no winning strategy is known (maybe you can discover one).

The game is called SIM. Six dots are drawn which would be the vertices of a hexagon. The two players use different colored pencils, or a pen and a pencil so that each player's lines can be distinguished from the others. The players take turns drawing a side or a diagonal of the hexagon. The first player to complete a triangle, with any three dots, *all in his color* loses.

How many line segments can be drawn?

How many different triangles are possible?

Notice the extensive use of geometric shapes in the construction of the Oakland Coliseum. *Courtesy Oakland Coliseum.*

## 18.  Quadrilaterals

We have already studied rather extensively one subset of the set of polygons—the triangle. We would now like to look at the set of the quadrilaterals. There are many different kinds of quadrilaterals that we classify, and many theorems about them. The parallelogram is especially useful, and we will be interested in ways of proving a figure is a parallelogram. A Venn diagram will be useful in visualizing the various types of quadrilaterals and their relationship to each other.

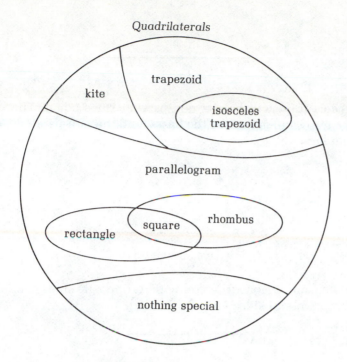

A **quadrilateral** is a polygon with four sides. (We will restrict our discussion to convex quadrilaterals.)

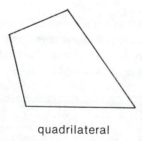

quadrilateral

A **kite** is a quadrilateral with two pairs of equal consecutive sides. Each pair must be a different length.

kite

A **trapezoid** is a quadrilateral with one, and only one, pair of sides parallel.

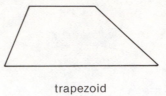

trapezoid

An **isosceles trapezoid** has its non-parallel sides equal. The parallel sides of a trapezoid are called the **bases,** and the non-parallel sides are called the **legs.**

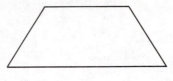

isosceles trapezoid

A **parallelogram** is a quadrilateral with its opposite sides parallel. This means both pairs of sides must be parallel.

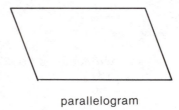

parallelogram

A **rectangle** is a parallelogram with one right angle. It is only necessary to know that one angle is a right angle, because we will see later that the properties of a parallelogram will guarantee that all of the angles will be right angles if any one angle is a right angle.

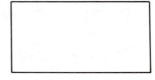

rectangle

A **rhombus** is a parallelogram with two adjacent sides equal. Again, the properties of a parallelogram will guarantee that all sides are equal if any adjacent pair are equal.

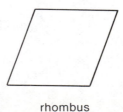

rhombus

If we ignore the fact that rectangles are a subset of the parallelograms we could define a rectangle as a quadrilateral with four right angles. Likewise, we could define a rhombus as a quadrilateral with four equal sides. A rectangle is an equiangular quadrilateral. A rhombus is an equilateral quadrilateral.

The **square** is unique in that it is both a rectangle and a rhombus. Consequently we can define it either as a rectangle with two adjacent sides equal, or as a rhombus with one right angle. Ignoring that it is a subset of rectangle and rhombus, it could be defined as a regular quadrilateral; that is, both equilateral and equiangular.

square

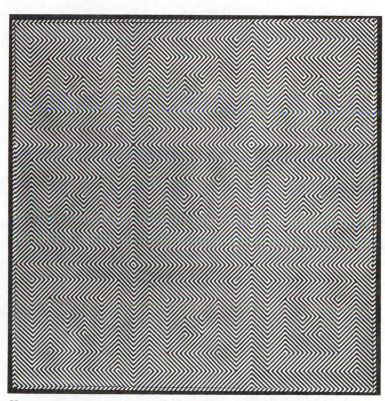

How many squares can you find in this painting? (*Square of Three— Yellow and Black,* by Reginald Neal) *Courtesy New Jersey State Museum Collection, Trenton; museum purchase.*

We would now like to examine some of the properties of these various figures. First, for the kite, only one outstanding property needs to be mentioned:

---

> ### *The diagonals of a kite are perpendicular to each other.*

Given: $ABCD$ is a kite

Prove: $BD \perp AC$

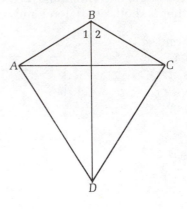

| | |
|---|---|
| 1. $ABCD$ is a kite | 1. Given |
| 2. $AB = BC$ and $AD = CD$ | 2. Def. of kite |
| 3. $BD = BD$ | 3. Reflexive |
| 4. $\triangle ABD \cong \triangle BCD$ | 4. SSS |
| 5. $\angle 1 = \angle 2$ | 5. CPCTE |
| 6. In $\triangle ABC$, $BD \perp AC$ | 6. Bis. of vertex $\angle$ of isos. $\triangle$ |

---

Notice that the converse to the theorem is not true. A quadrilateral can have diagonals that are perpendicular to each other and not be a kite.

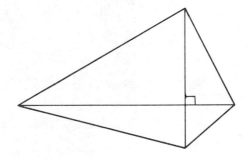

There are two special properties of the isosceles trapezoid with which we should be familiar:

---

## The base angles of an isosceles trapezoid are equal.

Given: Trapezoid $ABCD$ with $AB = CD$, $BC \parallel AD$

Prove: $\angle A = \angle D$

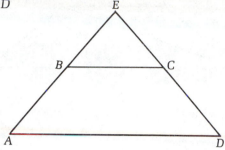

| | |
|---|---|
| **1.** $ABCD$ is a trapezoid with $BC \parallel AD$ | **1.** Given |
| **2.** Extend $AB$ and $DC$ to intersect at $E$ | **2.** Construction ($AB$ & $DC$ cannot be parallel by def. of trap.) |
| **3.** $\dfrac{AB}{BE} = \dfrac{CD}{CE}$ | **3.** $\parallel$ line divides sides prop. |
| **4.** $AB \cdot CE = BE \cdot CD$ | **4.** Prod. of means = prod. of ext. |
| **5.** $AB = CD$ | **5.** Given |
| **6.** $CE = BE$ | **6.** Division |
| **7.** $AE = DE$ | **7.** Addition |
| **8.** $\angle A = \angle D$ | **8.** Angles opposite = sides in $\triangle ADE$ |

---

### The diagonals of an isosceles trapezoid are equal.

Given:  Trapezoid $ABCD$ with $AB = CD$, $BC \parallel AD$

Prove:  $AC = BD$

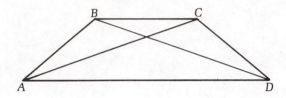

| 1. $ABCD$ is a trapezoid with $BC \parallel AD$ | 1. Given |
| 2. $AB = CD$ | 2. Given |
| 3. $\angle BAD = \angle CDA$ | 3. Base $\angle$s of isos. trap. are $=$ |
| 4. $AD = AD$ | 4. Reflexive |
| 5. $\triangle BAD \cong \triangle CDA$ | 5. SAS |
| 6. $BD = CA$ | 6. CPCTE |

---

Notice that the converse of the theorem is not true. A quadrilateral can have equal diagonals and yet not be an isosceles trapezoid.

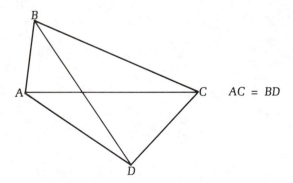

$AC = BD$

There are three very important properties of the parallelogram which we should prove:

| The opposite sides of a parallelogram are equal. |
| :---: |

Given: ABCD is a parallelogram

Prove: AB = CD, AD = BC

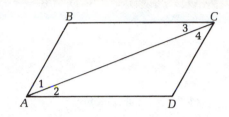

| | |
| --- | --- |
| **1.** Draw AC | **1.** Construction |
| **2.** AB ∥ CD | **2.** Def. of parallelogram |
| **3.** ∡1 = ∡4 | **3.** Alt. int. angles |
| **4.** AD ∥ BC | **4.** Def. of parallelogram |
| **5.** ∡3 = ∡2 | **5.** Alt. int. angles |
| **6.** AC = AC | **6.** Reflexive |
| **7.** △ABC ≅ △ADC | **7.** ASA |
| **8.** AB = CD and AD = BC | **8.** CPCTE |

The converse of this theorem is true and we will prove it shortly. The second important property of parallelograms is:

---

### The opposite angles of a parallelogram are equal.

Given: ABCD is a parallelogram

Prove: ∡BAD = ∡BCD, ∡B = ∡D

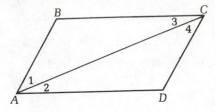

| | |
|---|---|
| 1. ABCD is a parallelogram | 1. Given |
| 2. AB = CD and AD = BC | 2. Opp. sides of a ▱ are = |
| 3. AC = AC | 3. Reflexive |
| 4. △ABC ≅ △CDA | 4. SSS |
| 5. ∡B = ∡D | 5. CPCTE |
| 6. ∡1 = ∡4 and ∡2 = ∡3 | 6. CPCTE |
| 7. ∡BAD = ∡BCD | 7. Addition |

---

The converse of this theorem is also true.
And now, the third major property of parallelograms:

---

### The diagonals of a parallelogram bisect each other.

Given: ABCD is a parallelogram

Prove: AO = OC and BO = OD

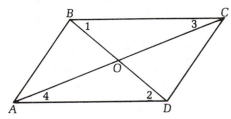

| | |
|---|---|
| 1. ABCD is a parallelogram | 1. Given |
| 2. BC ∥ AD | 2. Def. of parallelogram |
| 3. ∡1 = ∡2 and ∡3 = ∡4 | 3. Alt. int. ∡s |
| 4. AD = BC | 4. Opp. sides of a ▱ are = |
| 5. △BOC ≅ △DOA | 5. ASA |
| 6. AO = OC and BO = OD | 6. CPCTE |

We will also prove that the converse of this theorem is true. But first, there is one corollary that proves nicely here.

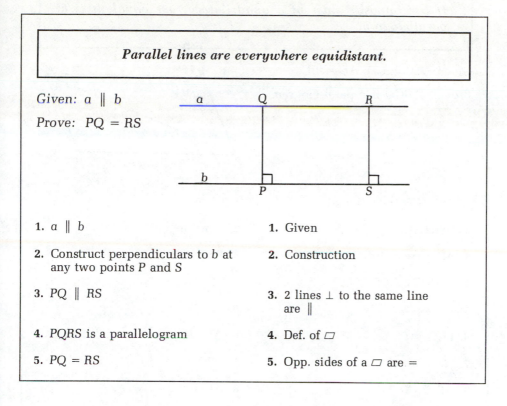

### Parallel lines are everywhere equidistant.

Given: $a \parallel b$

Prove: $PQ = RS$

| | |
|---|---|
| 1. $a \parallel b$ | 1. Given |
| 2. Construct perpendiculars to $b$ at any two points $P$ and $S$ | 2. Construction |
| 3. $PQ \parallel RS$ | 3. 2 lines $\perp$ to the same line are $\parallel$ |
| 4. $PQRS$ is a parallelogram | 4. Def. of $\square$ |
| 5. $PQ = RS$ | 5. Opp. sides of a $\square$ are $=$ |

Now to prove all those converses that you were promised! The definition of a parallelogram, the three theorems which are converses of the ones just proven, and one more theorem will give us five different ways of proving that a quadrilateral is a parallelogram. Since so many properties are true once a figure has been proven a parallelogram, and since the rectangle, rhombus, and square are defined in terms of the parallelogram, these methods become quite useful.

*If the opposite sides of a quadrilateral are equal, it is a parallelogram.*

Given: $AB = CD$, $AD = BC$

Prove: $ABCD$ is a parallelogram

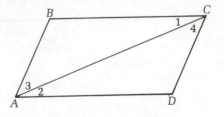

| | |
|---|---|
| **1.** $AB = CD$ and $AD = BC$ | **1.** Given |
| **2.** Draw $AC$ | **2.** Construction |
| **3.** $AC = AC$ | **3.** Reflexive |
| **4.** $\triangle ABC \cong \triangle CDA$ | **4.** SSS |
| **5.** $\angle 1 = \angle 2$ | **5.** CPCTE |
| **6.** $BC \parallel AD$ | **6.** Alt. int. $\angle$s = |
| **7.** $\angle 3 = \angle 4$ | **7.** CPCTE |
| **8.** $AB \parallel CD$ | **8.** Alt. int. $\angle$s = |
| **9.** $ABCD$ is a $\square$ | **9.** Def. of $\square$ |

In this bridge railing you should be able to find parallelograms, trapezoids, and triangles.

> **If the opposite angles of a quadrilateral are equal, it is a
> parallelogram.**

Given:  $\angle A = \angle C, \angle B = \angle D$

Prove:  $ABCD$ is a parallelogram

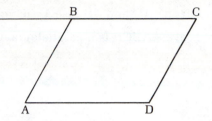

| | |
|---|---|
| **1.** $\angle A = \angle C$ and $\angle B = \angle D$ | **1.** Given |
| **2.** $\angle A + \angle B + \angle C + \angle D = 360$ | **2.** Sum of the angles of a polygon of 4 sides |
| **3.** $2\angle A + 2\angle B = 360$ | **3.** Substitution |
| **4.** $\angle A + \angle B = 180$ | **4.** Division |
| **5.** Extend $CB$ to $E$ | **5.** Construction |
| **6.** $\angle ABE + \angle B = 180$ | **6.** Straight angle |
| **7.** $\angle ABE = \angle A$ | **7.** Supplements of same angle |
| **8.** $AD \parallel BC$ | **8.** If alt. int. $\angle$s =, lines are $\parallel$ |

In like manner show $AB \parallel CD$

| | |
|---|---|
| **9.** $ABCD$ is a parallelogram | **9.** Def. of parallelogram |

*If the diagonals of a quadrilateral bisect each other, it is a parallelogram.*

Given: $AC$ and $BD$ bisect each other

Prove: $ABCD$ is a parallelogram

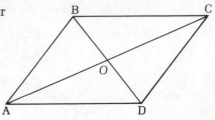

| | |
|---|---|
| **1.** $AC$ and $BD$ bisect each other | **1.** Given |
| **2.** $AO = OC$ and $BO = OD$ | **2.** Def. of bisect |
| **3.** $\angle BOC = \angle AOD$ | **3.** Vertical angles are = |
| **4.** $\triangle BOC \cong \triangle DOA$ | **4.** SAS |
| **5.** $BC = AD$ | **5.** CPCTE |
| **6.** $\angle AOB = \angle COD$ | **6.** (3) |
| **7.** $\triangle AOB \cong \triangle COD$ | **7.** SAS |
| **8.** $AB = CD$ | **8.** CPCTE |
| **9.** $ABCD$ is a parallelogram | **9.** Opp. sides are = |

There is one additional method of showing that a quadrilateral is a parallelogram that is sometimes convenient, as it deals with only one pair of sides.

> *If a quadrilateral has one pair of sides both equal and parallel, then it is a parallelogram.*

Given: $AB \parallel CD$ and $AB = CD$

Prove: $ABCD$ is a parallelogram

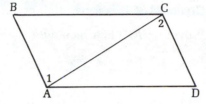

| | |
|---|---|
| 1. $AB \parallel CD$ | 1. Given |
| 2. $AB = CD$ | 2. Given |
| 3. Draw $AC$ | 3. Construction |
| 4. $\angle 1 = \angle 2$ | 4. Alt. int. angles |
| 5. $AC = AC$ | 5. Reflexive |
| 6. $\triangle ABC \cong \triangle CDA$ | 6. SAS |
| 7. $AD = BC$ | 7. CPCTE |
| 8. $ABCD$ is a $\square$ | 8. If the opp. sides of a quadrilateral are $=$, it is a parallelogram |

Now let us continue with the investigation of properties of different kinds of quadrilaterals. We have not yet discussed the properties of a rectangle, rhombus, or square. Recall that we have defined the rectangle as a parallelogram with a right angle. We can show that:

---

**If the diagonals of a parallelogram are equal, it is a rectangle.**

---

Given: *ABCD* is a parallelogram, *AC* = *BD*

Prove: *ABCD* is a rectangle

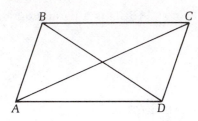

| | |
|---|---|
| **1.** *ABCD* is a parallelogram | **1.** Given |
| **2.** *BD* = *AC* | **2.** Given |
| **3.** *AB* = *CD* | **3.** Opp. sides of a $\square$ |
| **4.** *AD* = *AD* | **4.** Reflexive |
| **5.** $\triangle ABD \cong \triangle DCA$ | **5.** SSS |
| **6.** $\angle BAD = \angle CDA$ | **6.** CPCTE |
| **7.** $\angle BAD = \angle BCD$ | **7.** Opp. $\angle$s of a $\square$ |
| **8.** $\angle CDA = \angle ABC$ | **8.** Opp. $\angle$s of a $\square$ |
| **9.** $\angle BAD + \angle ABC + \angle BCD + \angle CDA = 360$ | **9.** Sum of the $\angle$s of a polygon of 4 sides |
| **10.** $4\angle BAD = 360$ | **10.** Substitution |
| **11.** $\angle BAD = 90$ | **11.** Division |
| **12.** *ABCD* is a rectangle | **12.** Def. of rectangle |

This theorem has a great many practical uses. If you are cutting some material, laying a foundation, making a picture frame, cutting a board, or in any manner trying to make a rectangle, measuring the opposite sides equal will not necessarily make a rectangle; it could be a parallelogram. However, if the diagonals measure the same length, then you can be assured that your figure is a rectangle. This technique is much used by construction workers. Of course, equal diagonals alone do not make a rectangle. The opposite sides must be equal, or the opposite sides parallel, or one pair of sides equal and parallel, or the figure must satisfy one of our other theorems in order to guarantee that the figure is a parallelogram. Then, if the diagonals are equal, it is a rectangle.

There is one method, other than the definition, of assuring that a figure is a rhombus:

> *If the diagonals of a quadrilateral are perpendicular bisectors of each other, then it is a rhombus.*

Given:  $AC$ and $BD$ are $\perp$ bisectors of each other

Prove:  $ABCD$ is a rhombus

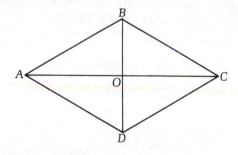

(Analysis: Since the only method we have to show that a figure is a rhombus is the definition, we will have to first show that the quadrilateral is a parallelogram, and then show that a pair of adjacent sides are equal.)

| | |
|---|---|
| 1. $AC$ and $BD$ are $\perp$ bis. of ea. other | 1. Given |
| 2. $ABCD$ is a parallelogram | 2. Diag. bis. each other |
| 3. $BD \perp AC$ | 3. Given |
| 4. $\angle BOC = \angle BOA$ | 4. Rt. angles are $=$ |
| 5. $AO = OC$ | 5. Def. of bisect |
| 6. $OB = OB$ | 6. Reflexive |
| 7. $\triangle BOC \cong \triangle BOA$ | 7. SAS |
| 8. $AB = BC$ | 8. CPCTE |
| 9. $ABCD$ is a rhombus | 9. Def. of rhombus |

Finally, we need to consider the square. Since it is both a rectangle and a rhombus, it will have the properties of both. The diagonals will be equal because it is a rectangle, and the diagonals will be perpendicular bisectors of each other since it is a rhombus. So, if a quadrilateral is a square, its diagonals are perpendicular bisectors of each other. We now prove that the converse is true:

---

> *If the diagonals of a quadrilateral are equal and perpendicular bisectors of each other, then it is a square.*

*Given:* $AC = BD$, $AC$ and $BD \perp$ bisectors of each other.

*Prove:* $ABCD$ is a square.

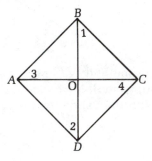

| | |
|---|---|
| 1. $AC = BD$ | 1. Given |
| 2. $AC$ and $BD \perp$ bis. of ea. other | 2. Given |
| 3. $AO = BO = CO = DO$ | 3. Halves of equals are $=$ |
| 4. $\angle AOB = \angle BOC = \angle COD = \angle DOA$ | 4. All rt. angles are $=$ |
| 5. $\triangle AOB \cong \triangle BOC \cong \triangle COD \cong \triangle DOA$ | 5. SAS |
| 6. $\angle 1 = \angle 2$ and $\angle 3 = \angle 4$ | 6. CPCTE |
| 7. $BC \parallel AD$ and $AB \parallel CD$ | 7. Alt. int. angles $=$ |
| 8. $ABCD$ is a parallelogram | 8. Def. of parallelogram |
| 9. $ABCD$ is a rectangle | 9. Parallelogram with $=$ diagonals |
| 10. $ABCD$ is a rhombus | 10. Diagonals bis. ea. other |
| 11. $ABCD$ is a square | 11. Def. of square (properties of both a rectangle and a rhombus) |

We should probably summarize, for easy reference, the methods we have developed for proving that a quadrilateral is a parallelogram. A quadrilateral is a parallelogram if:

the opposite sides are equal;
the opposite angles are equal;
the diagonals bisect each other;
one pair of sides is both equal and parallel; or
the opposite sides are parallel (Definition).

It is also useful to be able to classify the various quadrilaterals by the special properties of their diagonals.

In order to better organize the information on diagonals, let us use this notation. "If $a$ then $b$" can be written as $a \Rightarrow b$. The converse "if $b$ then $a$" can be written as $a \Leftarrow b$. If both are true, we will write $a \Leftrightarrow b$, which is read "if and only if".

| Type of quadrilateral | | Diagonals are |
|---|---|---|
| Kite | $\Rightarrow$ | perpendicular |
| Trapezoid | $\Rightarrow$ | ? |
| Isosceles trapezoid | $\Rightarrow$ | equal |
| Parallelogram | $\Leftrightarrow$ | bisect each other |
| Rectangle | $\Leftrightarrow$ | bisect each other, equal |
| Rhombus | $\Leftrightarrow$ | perpendicular bisectors of each other |
| Square | $\Leftrightarrow$ | perpendicular bisectors of each other, equal |

*1–17. True–false*

1. A quadrilateral with four right angles is a square.

2. The diagonals of a rectangle are perpendicular to each other.

3. The base angles of an isosceles trapezoid are equal.

4. If the diagonals of a quadrilateral are equal, it is a rectangle.

5. A diagonal of a parallelogram divides it into two congruent triangles.

6. The diagonals of a trapezoid are equal.

7. Every square is a parallelogram.

8. Every equilateral quadrilateral is equiangular.

9. The diagonals of a parallelogram are equal.

10. The diagonals of a rhombus are equal.

11. The diagonals of a rhombus bisect the angles of the rhombus.

12. If the diagonals of a quadrilateral are perpendicular to each other, the figure is a rhombus.

13. The diagonals of a parallelogram bisect each other.

14. If two sides of a quadrilateral are equal and the other two sides are parallel, the figure is a parallelogram.

15. The diagonals of a square are equal and perpendicular to each other.

16. If a pair of adjacent angles of a quadrilateral are supplementary, the figure is a parallelogram.

17. If a perpendicular is drawn to the diagonal of a square from any point on one of its sides, an isosceles right triangle is formed.

18. Prove: If a quadrilateral is a rectangle, all the angles are right angles.

19. Prove: If a quadrilateral is a rectangle, the diagonals are equal. (You may use the results of the previous exercise.)

20. Prove: If a quadrilateral is a rhombus, all its sides are equal.

21. Given: ABCD is a parallelogram
       $CE = AD$

    Prove: $\angle A = \angle E$

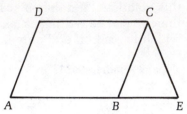

22. Given: ABCD is a parallelogram
       BCEF is a parallelogram

    Prove: AFED is a parallelogram

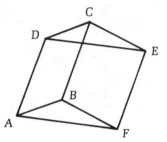

23. Given: ABCD is a trapezoid
       $AD = CD$

    Prove: AC bisects $\angle A$

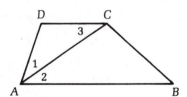

24. Given: ABCD is a parallelogram
       $BF = DH$
       $AE = CG$

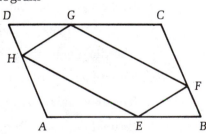

    Prove: EFGH is a parallelogram

25. Prove: The opposite angles of an isosceles trapezoid are supplementary.

26. In parallelogram PQRS, $\angle P$ is 43° greater than $\angle Q$. How large is $\angle S$?

27. Prove: The quadrilateral formed by joining the midpoints of the sides of any quadrilateral is a parallelogram.

28. *Given:* ABCD is a parallelogram
    CDFE is a parallelogram

    *Prove:* ABEF is a parallelogram

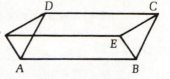

29. *Prove:* The quadrilateral formed by joining the midpoints of the sides of a rectangle is a rhombus.

30. *Prove:* A diagonal of a rhombus bisects its angles.

*31. *Prove:* If the midpoints of three sides of a rhombus are joined, the resulting triangle is a right triangle.

*32. *Prove:* If the four sides of a square are extended equal distances and their endpoints joined, the resulting figure is also a square.

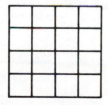

*33. *Prove:* The bisectors of the angles of a rectangle form a square.

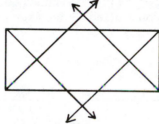

34. How many squares, of any size, are there in this grid?

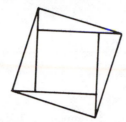

35. How many squares are there in a grid 6 units on a side?

*36. How many squares, of any size, are there in a grid N units on a side?

*37. How many cubes, of any size, are there in this three-dimensional grid? (A cube has all edges the same length.)

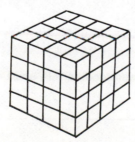

*38. How many cubes of any size can be found in a cube six units on a side?

**39. How many cubes, of any size, can be found in a cube N units on a side?

*B.C. by permission of Johnny Hart and Field Enterprises, Inc.*

## 19. Quadrilateral constructions

The construction of quadrilaterals, where only certain minimum information is given, makes for problems that challenge the ingenuity. The key to most constructions of this type is knowing the properties of the quadrilateral to be constructed. Some examples:

**Construct a rhombus, given the diagonals.**

**Since we know that the diagonals of a rhombus are perpendicular bisectors of each other, construct the perpendicular bisector of $d_1$ and mark off $\frac{1}{2}d_2$ (found by bisecting $d_2$) on each side.**

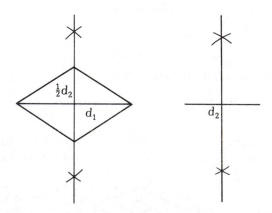

**Construct a kite, given the shorter side, $AB$, and both diagonals.**

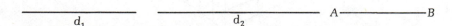

Since the diagonals of a kite are perpendicular (and one is the bisector of the other) construct a perpendicular bisector to $d_1$; from one end of $d_1$ mark an arc equal to $AB$ cutting the perpendicular. From that intersection mark off $d_2$ and then complete the kite.

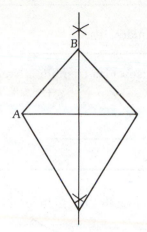

1. Construct a square, given a side: ────────────

2. Construct a rhombus, given that one angle is 60°, and the perimeter: ──────────────────────

3. Construct a parallelogram, given one side: ──────────────

4. Construct a rectangle, given one side: ──────────────

5. Construct an isosceles trapezoid, given
the longer base: ────────────────────────
one of the equal legs; ──────────────
a base angle;

6. Construct a rhombus, given
a side: ──────────────
the shorter diagonal; ──────────────────

7. Construct a parallelogram, given
the shorter diagonal: ──────────────────
the longer diagonal; ────────────────────────────
the angle of intersection of the diagonals;

8. Construct a parallelogram, given
the shorter diagonal: ────────────────────
the longer diagonal; ────────────────────────
a side; ──────────────

9. Construct an isosceles trapezoid, given
the longer base: _____
a diagonal; _____
one of the equal legs; _____

10. Construct a square, given
a diagonal: _____

11. Construct an isosceles triangle, given the base $AB$ and $\angle A$.

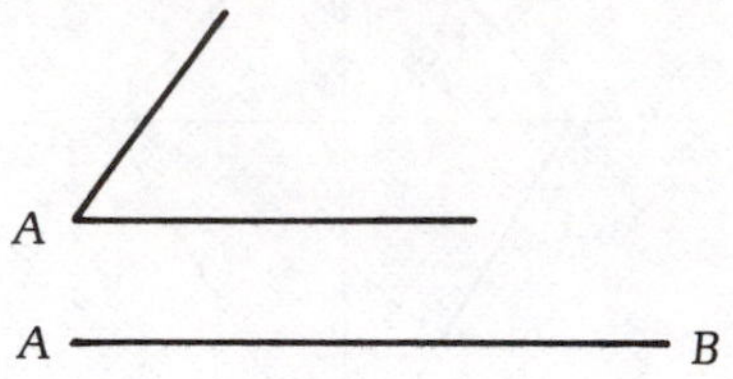

A

A ———————————— B

12. Construct an isosceles triangle, given the base $AB$ and $\angle C$.

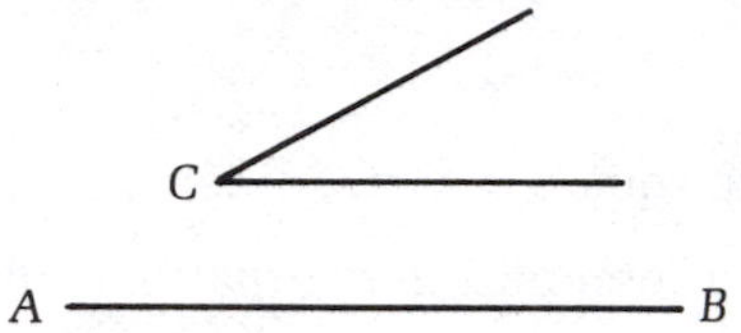

C

A ———————————— B

13. Construct an acute triangle $ABC$, given $AB$, $BC$, and $\angle C$.

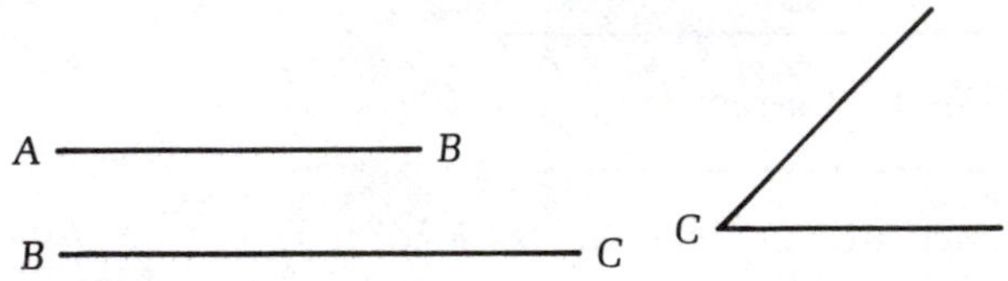

A ——————— B

B ——————— C    C

14. Construct an obtuse triangle $ABC$, using the given of exercise 13.

15. Construct a regular hexagon, given the apothem (the line from the center perpendicular to a side). ____$a$____

16. Construct an isosceles trapezoid, given
longer base: _____
shorter base; _____
leg; _____

17. Construct a rectangle, given
short side: _____
diagonal; _____

18. Construct a parallelogram, given
an angle:

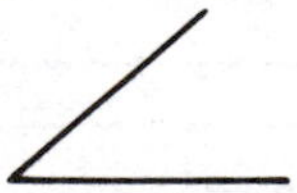

longer side; _____
longer diagonal; _____

*19–23. If you like puzzles, you might try arranging the five pieces of this square:

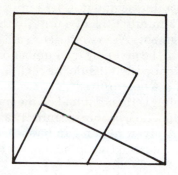

into these shapes:

19.

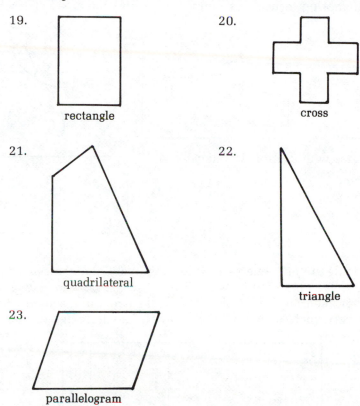

rectangle

20.

cross

21.

quadrilateral

22.

triangle

23.

parallelogram

(Obviously, the five shapes are not drawn to scale.)

*24. Find the intersection, P, of the two lines AB and CD, when there is an obstruction, X, that will not permit the use of a straightedge to extend the line.

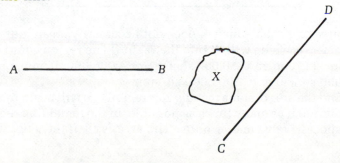

## 20. Areas of plane figures

When measuring a line segment, we give its length as the number of one-unit segments, and parts of unit segments, that will fit on the given line segment. To say a line segment is 1.73 units long means 1 unit, 7 tenths of a unit, and 3 hundredths of a unit (to the nearest hundredth) will fit on the line segment. We will measure area in the same manner. The area of a region will be given by the number of one unit squares and smaller unit squares that will fit inside the region. The smaller units are combined as parts of a whole unit.

The **area** of a region is then defined as **the number of unit squares that will fit in the region.** It is, of course, entirely arbitrary as to what size unit square is used. A given region can be measured in square miles, square centimeters, hectares (10,000 square meters), or square zonks (your own choice).

There are, however, some obvious difficulties that we encounter when attempting to use this definition to find the areas of irregular figures:

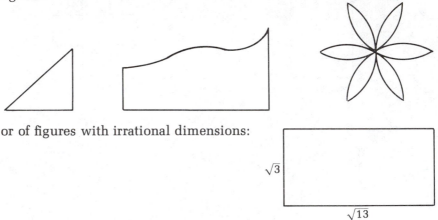

or of figures with irrational dimensions:

$\sqrt{3}$

$\sqrt{13}$

It is difficult to "fit" squares in these regions. Calculus is needed to properly solve these difficulties. The concept used to solve these "fitting" problems is the use of smaller and smaller unit squares. As we use very, very small squares, we can almost fill up the region in question.

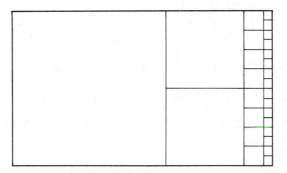

An area of 4 square meters does not necessarily mean four squares each 1 meter on a side will fit into the region. The region could contain 40,000 sq. cm (1 sq. m = 10,000 sq. cm), or 4,000,000 sq. mm, or 2 sq. m and 10,000 sq. cm and 1,000,000 sq. mm.

To find the area of some rectangular region, we will begin by fitting a given unit into it as many times as possible. In the remaining region, we will fit successively smaller units. The area is then the sum of all the units.

Suppose a given rectangular region is 3 × 1.73 units.

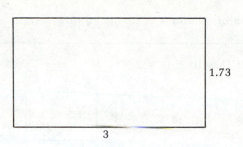

We can fit 3 unit squares, 210 squares each $\frac{1}{100}$ of a unit, and 900 (count them) squares each $\frac{1}{10\,000}$ of a unit. When this is added we get $3 + \frac{210}{100}$ $+ \frac{900}{10\,000} = 3 + 2.1 + .09 = 5.19$ square units. Notice that the length times the width $= 3 \times 1.73 = 5.19$ square units.

If neither the length nor the width are nice integers, the process becomes more complicated, but still yields the same results (try your own example). Since all the preceding is inductive, and not a proof, we make the following *definition:*

---

*The area of a rectangular region is its length times its width.*
**(A = lw)**

---

With this result, we can now examine many other figures: the square, parallelogram, triangle, rhombus, trapezoid, and polygon. In this section, we will restrict ourselves to finding the areas of some of the more common figures that have straight line boundaries.

> ### *The area of a square is a side squared.* ($A = s^2$)

Given: Square $ABCD$ with side $s$

Prove: $A = s^2$

| | |
|---|---|
| **1.** $ABCD$ is a square | **1.** Given |
| **2.** $AB = BC = s$ | **2.** Adj. sides of a square are $=$ |
| **3.** $ABCD$ is a rectangle | **3.** If it is a sq. it is a rect. |
| **4.** $A = s \cdot s = s^2$ | **4.** Area of rect. $= lw$ |

Notice the architect's extensive use of squares and rectangles. *Stanford Medical Center.*

> *The area of a parallelogram is its base times its height.* **(A = bh)**

Given: Parallelogram *ABCD*

Prove: $A_{ABCD} = bh$

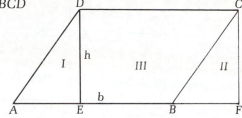

**NOTE:** We will designate the area of a figure such as a parallelogram *ABCD* as $A_{ABCD}$.

(Analysis: The plan is to show that the area of I equals the area of II using congruent triangles. Then "move" I over to II, giving us the area of a rectangle—which we already have a formula for.)

| | |
|---|---|
| 1. Draw *CF* ⊥ to *AB* produced | 1. Construction of a ⊥ to a line from a pt. |
| 2. *ABCD* is a parallelogram | 2. Given |
| 3. *DE* ∥ *CF* | 3. Lines ⊥ to same line are ∥ |
| 4. *DEFC* is a rectangle | 4. Opp. sides ∥ with a rt. angle |
| 5. $A_{DECF} = (DE)(EF)$ | 5. $A_{rect.}$ = length × width |
| 6. *AD* = *BC* | 6. Opp. sides of parallelogram |
| 7. ∢*A* = ∢*CBF* | 7. Corresponding angles |
| 8. ∢*ADE* = ∢*BCF* | 8. If 2∢s = 2∢s, 3rd ∢s of a △ are = |
| 9. △*ADE* ≅ △*BCF* | 9. ASA |
| 10. *AE* = *BF* | 10. CPCTE |
| 11. *BE* = *BE* | 11. Reflexive |
| 12. *AB* = *EF* | 12. Addition |
| 13. $A_{DEFC} = (DE)(AB)$ | 13. Substitution |
| 14. $A_{I} = A_{II}$ | 14. Congruent |
| 15. $A_{III} = A_{III}$ | 15. Reflexive |
| 16. $A_{ABCD} = A_{DEFC}$ | 16. Addition |
| 17. $A_{ABCD} = (DE)(AB) = bh$ | 17. Substitution |

> *The area of a triangle is $\frac{1}{2}$ its base times its height.* $(A_{ABC} = \frac{1}{2}bh)$

Given: Triangle $ABC$

Prove: $A_{ABC} = \frac{1}{2}bh$

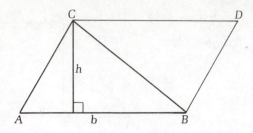

| | |
|---|---|
| **1.** Draw $BD \parallel AC$, $CD \parallel AB$ | **1.** Construction |
| **2.** $ABDC$ is a parallelogram | **2.** Opp. sides parallel |
| **3.** $AB = DC$, $AC = BD$ | **3.** Opp. sides of a parallelogram are $=$ |
| **4.** $BC = BC$ | **4.** Reflexive |
| **5.** $\triangle ABC \cong \triangle DCB$ | **5.** SSS |
| **6.** $\triangle ABC + \triangle DCB = ABDC$ | **6.** The whole of a quantity is equal to the sum of its parts |
| **7.** $2\triangle ABC = ABDC$ | **7.** Substitution |
| **8.** $\triangle ABC = \frac{1}{2}ABDC$ | **8.** Division |
| **9.** $A_{ABDC} = bh$ | **9.** Area of a parallelogram |
| **10.** $A_{ABC} = \frac{1}{2}bh$ | **10.** Substitution |

This is a most important theorem and it has a great many corollaries.

The corollaries which we will find most useful, especially for computations, will now be proven.

---

**Triangles with equal bases and equal heights are equal.**

Given: Triangles $ABC$ and $DEF$, both with base $b$ and height $h$

Prove: $\triangle ABC = \triangle DEF$

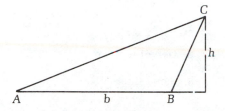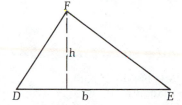

**NOTE:** Since "=" means "the same size," when we say $\triangle ABC = \triangle DEF$ it means the same as $A_{ABC} = A_{DEF}$.

1. $\triangle ABC = \frac{1}{2}bh$       1. Area of a triangle

2. $\triangle DEF = \frac{1}{2}bh$       2. Area of a triangle

3. $\triangle ABC = \triangle DEF$       3. Substitution

---

This is an interesting corollary, in that it says nothing about the shape of the triangle. The area depends only upon the base and height. To further illustrate this theorem, consider these triangles that have the same base, and their altitudes along a line parallel to the base. They will all have the same area!

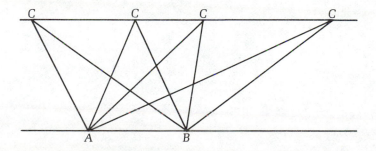

All $\triangle$s $ABC$ are equal!

A median is a line from a vertex to the midpoint of the opposite side.

---

### The median of a triangle divides it into two equal triangles.

Given: *ABC* with median *CD*

Prove: $\triangle ADC = \triangle BDC$

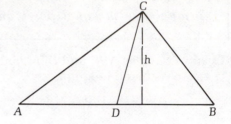

| | |
|---|---|
| **1.** *CD* is a median | **1.** Given |
| **2.** *AD = BD* | **2.** Def. of median |
| **3.** *h* is height of both triangles | **3.** Same line |
| **4.** $\triangle ADC = \triangle BDC$ | **4.** Equal bases and equal heights |

---

Again, notice that the shape of the two triangles may be different, with one acute, and the other right or obtuse, but the areas will be the same.

> *In a right triangle the product of the legs equals the product of the hypotenuse and the altitude to the hypotenuse.*

*Given:* Right triangle $ABC$ with altitude to the hypotenuse $CD$

*Prove:* $(AC)(BC) = (AB)(CD)$

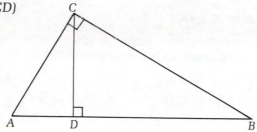

| | |
|---|---|
| **1.** $\triangle ABC$ is a rt. triangle | **1.** Given |
| **2.** $BC$ is alt. to base $AC$ | **2.** Def. of altitude |
| **3.** $A_{ABC} = \frac{1}{2}(AC)(BC)$ | **3.** Area of a triangle |
| **4.** $CD$ is alt. to $AB$ | **4.** Given |
| **5.** $A_{ABC} = \frac{1}{2}(AB)(CD)$ | **5.** Area of a triangle |
| **6.** $\frac{1}{2}(AC)(BC) = \frac{1}{2}(AB)(CD)$ | **6.** Substitution |
| **7.** $(AC)(BC) = (AB)(CD)$ | **7.** Multiplication |

As you might suspect, this corollary is very useful in solving numerical relationships. The same results can usually be found, however, by using similar triangles. As a matter of fact, this corollary can be proven using similar triangles:

| | |
|---|---|
| **1.** $\triangle ABC$ is a rt. triangle | **1.** Given |
| **2.** $CD$ is alt. to hypotenuse | **2.** Given |
| **3.** $\triangle ABC \sim \triangle ACD$ | **3.** Alt. to hyp. of rt. $\triangle$ makes $\sim \triangle$s |
| **4.** $\dfrac{AC}{CD} = \dfrac{AB}{BC}$ | **4.** Corr. parts of $\sim \triangle$s are prop. |
| **5.** $(AC)(BC) = (AB)(CD)$ | **5.** Prod. of means = prod. of extremes |

> *The area of an equilateral triangle with side s is $\dfrac{s^2\sqrt{3}}{4}$.*

Given: Equilateral triangle $ABC$

Prove: $A_{ABC} = \dfrac{s^2\sqrt{3}}{4}$

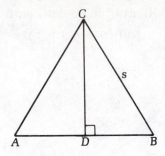

| | |
|---|---|
| 1. $\triangle ABC$ is equilateral | 1. Given |
| 2. $\measuredangle B = 60$ | 2. Each angle of equilateral triangle = 60 |
| 3. Draw $CD \perp AB$ | 3. Construction |
| 4. $CD = \dfrac{s\sqrt{3}}{2}$ | 4. Leg of 30-60-90 triangle |
| 5. $A_{ABC} = \frac{1}{2}(AB)(CD)$ | 5. Area of triangle |
| 6. $A_{ABC} = \frac{1}{2}(s)\left(\dfrac{s\sqrt{3}}{2}\right)$ | 6. Substitution |
| 7. $A_{ABC} = \dfrac{s^2\sqrt{3}}{4}$ | 7. Substitution |

> *The area of a triangle is one-half its perimeter times the radius of its inscribed circle. (A = ½pr)*

Given: △ABC with inscribed circle O with radius r

Prove: $A_{ABC} = \frac{1}{2}pr$

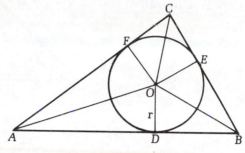

| | |
|---|---|
| **1.** Draw OA, OB, OC | **1.** Construction |
| **2.** Draw OD, OE, OF | **2.** Construction |
| **3.** OD ⊥ AB, OE ⊥ BC, OF ⊥ AC | **3.** A radius is ⊥ to a tangent at the point of contact. (Proved in the next chapter) |
| **4.** $A_{ABC} = A_{ABO} + A_{BCO} + A_{ACO}$ | **4.** Whole of a quant. = sum of its parts |
| **5.** $A_{ABO} = \frac{1}{2}r(AB)$, $A_{BCO} = \frac{1}{2}r(BC)$, $A_{ACO} = \frac{1}{2}r(AC)$ | **5.** Area of a triangle |
| **6.** $A_{ABC} = \frac{1}{2}r(AB) + \frac{1}{2}r(BC) + \frac{1}{2}r(AC)$ | **6.** Addition |
| **7.** $A_{ABC} = \frac{1}{2}r(AB + BC + AC)$ $= \frac{1}{2}rp$ | **7.** Substitution |

Did you notice that, in this theorem, we violated our rule about what constitutes acceptable reasons? We used a reason which was not a definition, postulate, axiom, or previously proven theorem. Reason (3) is a theorem from the next chapter. This theorem is usually proven after the chapter on circles, but to organize the material better, it was included here. The same concession will be made for the last theorem in this chapter, about the area of a regular polygon.

The area of a rhombus can be figured by the product of its base and height. This is because it is a parallelogram. But since the diagonals of a rhombus are perpendicular to each other, there is another way to calculate the area of a rhombus.

> **The area of a rhombus is one-half the product of the diagonals.**
> $(A = \frac{1}{2}d_1 d_2)$

Given: $ABCD$ is a rhombus

Prove: $A_{ABCD} = \frac{1}{2}(AC)(BD)$

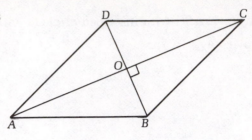

1. $ABCD$ is a rhombus
2. $AC \perp BD$
3. $A_{ABC} = \frac{1}{2}(AC)(OB)$
4. $A_{ADC} = \frac{1}{2}(AC)(OD)$
5. $A_{ABCD} = \frac{1}{2}(AC)(OB) + \frac{1}{2}(AC)(OD)$
6. $A_{ABCD} = \frac{1}{2}(AC)(OB + OD)$
7. $A_{ABCD} = \frac{1}{2}(AC)(BD)$

1. Given
2. Diagonals of a rhombus are $\perp$
3. Area of a triangle
4. Area of a triangle
5. The whole is = to the sum of its parts
6. Substitution (algebraic distribution)
7. Substitution

The theorem that we will look at now is rather amazing. It allows us to find the area of a triangle without knowing a height. The area is given in terms of the sides only. This is useful, since it is sometimes inconvenient or impossible in an actual situation to measure a height. It makes use of the variable $s$, which will represent the semiperimeter. If the sides are $a$, $b$, and $c$, then the perimeter is $a + b + c$, and the semiperimeter $s = \dfrac{a + b + c}{2}$. This theorem is attributed to Heron of Alexandria (about 250AD) and is called Heron's Semiperimeter Formula. The algebraic manipulations in the proof get rather sticky, and the student should not be expected to work through them. For the ambitious and competent algebraist, the proof is given in Appendix V.

> **The area of any triangle is** $\sqrt{s(s - a)(s - b)(s - c)}$ **where $s$ is the semiperimeter.**

As an illustration of the use of the theorem, find the area of a triangle with sides of 5, 7, and 9.

$$s = \frac{5 + 7 + 9}{2} = \frac{21}{2}$$

$$A = \sqrt{\frac{21}{2}\left(\frac{21}{2} - 5\right)\left(\frac{21}{2} - 7\right)\left(\frac{21}{2} - 9\right)} = \sqrt{\frac{21}{2}\left(\frac{11}{2}\right)\left(\frac{7}{2}\right)\left(\frac{3}{2}\right)}$$

$$A = \sqrt{\frac{7^2 \cdot 3^2 \cdot 11}{4^2}} = \frac{7 \cdot 3 \sqrt{11}}{4} = \frac{21\sqrt{11}}{4}$$

Next, we wish to find a formula to calculate the area of a trapezoid. There are *many* different proofs. We can add auxiliary lines to form a rectangle, two (or more) triangles, a parallelogram, a large triangle, or combinations of these. You should use this theorem to demonstrate to yourself your geometric competence. Find a method, other than the one given below, to prove this theorem. In many of the theorems we have presented, the student has not been expected to discover a proof on his own. This is one theorem for which this is not true. Perhaps you can find a more efficient method than the following:

---

*The area of a trapezoid is one-half the product of its height and the sum of its bases.* $[A = \frac{1}{2}h(b_1 + b_2)]$

---

Given: Trapezoid $ABCD$ with $AB \parallel CD$
Height $h$

Prove: $A_{ABCD} = \frac{1}{2}h(AB + CD)$

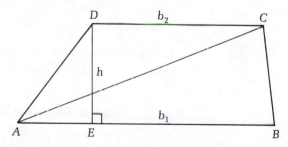

1. Draw $AC$
          1. Construction

2. Draw $DE \perp AB$
          2. Construction

3. $ABCD$ is trapezoid with $AB \parallel CD$
          3. Given

4. $DE \perp CD$
          4. $\perp$ to $\parallel$ lines

5. $A_{ABC} = \frac{1}{2}h(AB)$
          5. Area of a triangle

6. $A_{ACD} = \frac{1}{2}h(CD)$
          6. Area of a triangle

7. $A_{ABCD} = \frac{1}{2}h(AB) + \frac{1}{2}h(CD)$
          7. The whole is = to the sum of its parts

8. $A_{ABCD} = \frac{1}{2}h(AB + CD)$
          8. Substitution

There is one more figure for which we wish to develop a formula, the regular polygon. In a regular polygon, by drawing lines from its center to each vertex, you form congruent triangles. Since there are the same number of triangles as there are sides, the area of the polygon is n (the number of sides) times the area of one of the triangles. The line from the center of a regular polygon to a vertex is called a **radius.** It is the radius of the **circumscribed** circle. The line from the center perpendicular to a side is called the **apothem.** It is the radius of the **inscribed** circle. The **perimeter** is equal to n times the length of a side.

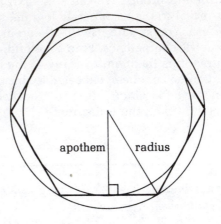

> *The area of a regular polygon is one-half its apothem times its perimeter. (A = $\frac{1}{2}ap$)*

Given:  Polygon ABCDE . . .
       with apothem $a$

Prove:  $A_{\text{polygon}} = \frac{1}{2}ap$

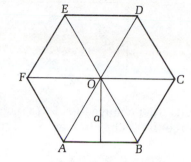

We will illustrate the theorem with a regular hexagon, although the proof does not depend upon the number of sides of the polygon.

| | |
|---|---|
| 1. Draw $OA$, $OB$, $OC$, $OD$, etc., forming $n$ triangles | 1. Construction |
| 2. $AB = BC = CD =$ etc. | 2. Polygon is regular |
| 3. $OA = OB = OC =$ etc. | 3. All radii of the same circle are equal (From def. of a circle, in the next chapter) |
| 4. $\triangle ABO \cong \triangle BCO \cong \triangle CDO \cong$ etc. | 4. SSS |
| 5. $A_{\text{polygon}} = n(A_{ABO})$ | 5. Whole = sum of its parts |
| 6. $a \perp AB$ | 6. Definition of an apothem |
| 7. $A_{ABO} = \frac{1}{2}(AB)a$ | 7. Area of a triangle |
| 8. $A_{\text{polygon}} = \frac{1}{2}an(Ab)$ | 8. Substitution |
| 9. $n(AB) = p$ | 9. Whole = sum of its parts |
| 10. $A_{\text{polygon}} = \frac{1}{2}ap$ | 10. Substitution |

Before we illustrate the theorems of this section with some examples, we should say something about dimensions. It is traditional, but not necessary, to compute area in square units and volume in cubic units. In order to have square units, the dimensions must have the same denominations. If a figure has a length in feet and a width in inches, and we multiply length times width without making a conversion, we get an area measured in feet-inches. Most of the time, this will be meaningless. (There are exceptions—some materials are measured in running measure and lumber is measured in board feet, which is 1 ft. by 1 ft. by 1 in.) In many of the exercises, no dimensions will be given. When this is the case, you are to assume that all measurements have the same dimension, and the resulting area will be square units of that dimension. If measurements are given in different dimensions, be certain to change to a common dimension and give your answer in square units.

**Find the area of a rectangle with length of 7 ft. and width of 20 in.**

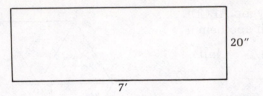

You can either change the feet to inches: $A = (7 \times 12)(20) = 1680$ sq. in. or change the inches to feet: $A = (7)(\frac{20}{12}) = \frac{35}{3} = 11\frac{2}{3}$ sq. ft.

**Find the area of the trapezoid with height of 17 mm and bases of 5 cm and 7 cm.**

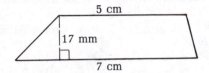

Choosing to give the answer in cm²: $A = \frac{1}{2}h(b_1 + b_2) = \frac{1}{2}(1.7)(5 + 7) = \frac{1}{2}(1.7)(12) = (1.7)(6) = 10.2$ cm²

**Find the area of the illustrated triangle.**

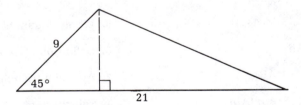

Using the ratio of the sides of an isosceles right triangle, $h = \dfrac{9}{\sqrt{2}}$

$A = \frac{1}{2}bh = \frac{1}{2}(21)\left(\dfrac{9}{\sqrt{2}}\right) = \dfrac{189\sqrt{2}}{4}$ square units

**Find the area of the equilateral triangle with side of 5.**

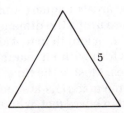

Using the formula $A = \dfrac{s^2\sqrt{3}}{4}$, $A = \dfrac{25\sqrt{3}}{4}$ square units

**Find the radius of the circle inscribed in a triangle with sides of 3, 5, and 7.**

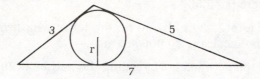

Here we will make use of two formulas. First find the area by use of Heron's Semiperimeter Formula, then use the area and the perimeter in the formula $A = \frac{1}{2}rp$ to find the radius.

$$A = \sqrt{s(s-a)(s-b)(s-c)} = \sqrt{\left(\frac{15}{2}\right)\left(\frac{9}{2}\right)\left(\frac{5}{2}\right)\left(\frac{1}{2}\right)} = \sqrt{\frac{(5^2)(9)(3)}{2^4}}$$

$$= \frac{15}{4}\sqrt{3}.$$

So, $\frac{15}{4}\sqrt{3} = \frac{1}{2}rp$, $\frac{15}{4}\sqrt{3} = \frac{15}{2}r$, $r = \frac{\sqrt{3}}{2}$.

**Find the area of the rhombus with an angle of 60°, and half the longest diagonal 5 units.**

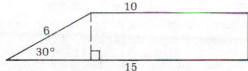

You should be able to determine that the diagonals bisect the angles, so that each triangle is a 30-60-90 triangle. Since the longest leg is 5, the shorter leg is $\frac{5}{\sqrt{3}}$. Then $A = \frac{1}{2}d_1d_2 = \frac{1}{2}(10)\left(\frac{10}{\sqrt{3}}\right) = \frac{50\sqrt{3}}{3}$ square units.

**Find the area of a trapezoid with a leg of 6, adjacent base angle of 30°, and bases of 10 and 15.**

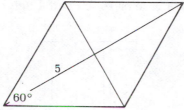

Using the ratio of the sides of a 30-60-90 triangle, the height will be 3.
$A = \frac{1}{2}h(b_1 + b_2) = \frac{1}{2}(3)(10 + 15) = \frac{75}{2}$ square units

**Find the area of a regular hexagon with radius of 8.**

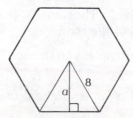

You should be able to determine that a side and two radii form an equilateral triangle. Its altitude, which is the apothem, will be $4\sqrt{3}$. So $A = \frac{1}{2}ap = \frac{1}{2}(4\sqrt{3})(48) = 96\sqrt{3}$ square units.

1. If a lot is 60 ft. by 120 ft., the lot is what fractional part of an acre? (One acre is 43,560 sq. ft.)

2. If a lot is 20 m by 40 m, the lot is what fractional part of a hectare? (One hectare is 10,000 sq. m.)

3. If the perimeter of a rectangular garden is 20 m and its area is 24 sq. m, find its dimensions.

4. Find the area of the cross section of a steel L-beam made of material 1 cm wide, and having a base of 8 cm and a height of 12 cm.

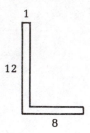

5. Find the cross sectional area of an I-beam made of material 2 cm wide, and having a base of 16 cm and a height of 36 cm.

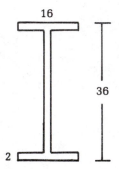

6. Find the area of a parallelogram with a base of 11 cm and altitude of 8 cm.

7. The area of a parallelogram is 72 sq. cm and the base is $9\sqrt{2}$ cm. Find the altitude of the parallelogram.

8. Two adjacent sides of a parallelogram are 8 cm and 14 cm and they include an angle of 60°. Find the area of the parallelogram.

9. Find the area of a parallelogram if two adjacent sides are 10 cm and 16 cm, and the two angles of the parallelogram are in the ratio of 3:1.

10. The perimeter of a rhombus is 48 cm and one angle is 30°. Find the area of the rhombus.

11. The area of a triangle is 80 sq. cm. The base is 16 cm. Find the altitude.

12. Find the area of an equilateral triangle with an altitude of $5\sqrt{3}$.

13. Find the area of an equilateral triangle, each side of which is 3 cm.

14. Find the side of an equilateral triangle whose area is $6\sqrt{3}$ sq. cm.

15. The sides of a triangle are 6 cm and 10 cm and the included angle is 30°. Find the area.

16. Two sides of a triangle are 4 cm and 6 cm and the included angle is 45°. Find the area.

17. The base of an isosceles triangle is 20 cm and a leg is 26 cm. Find the area.

18. Find the area of a triangle whose perimeter is 26 cm and the radius of the inscribed circle is 2 cm.

19. Find the area of a triangle that has sides of 3 m, 5 m, and 6 m.

20. Find the area of an isosceles right triangle whose hypotenuse is 6 cm.

21. If the area of an isosceles right triangle is 16 sq. cm, find the length of the triangle's hypotenuse.

22. One diagonal of a rhombus is 12 cm and the area is 96 sq. cm. Find the other diagonal.

23. A square with an area of 81 sq. cm has the same perimeter as an equilateral triangle. Find the area of the equilateral triangle.

24. The diagonals of a parallelogram are 12 cm and 18 cm and intersect at an angle of 30°. Find the area of the parallelogram.

25. Find the area of a trapezoid with bases of 10 cm and 12 cm and with an altitude of 5 cm.

26. The bases of an isosceles trapezoid are 9 cm and 15 cm and each leg is 5 cm. Find the area of the trapezoid.

27. The shorter base of an isosceles trapezoid is 7 cm, each leg is $6\sqrt{2}$ cm, and each base angle is 45°. Find the area of the trapezoid.

28. Find the area of an isosceles trapezoid, when the shorter base is 12 cm, the altitude is $4\sqrt{3}$ cm, and one of the angles is twice the other.

29. The bases of an isosceles trapezoid are 9 cm and 15 cm and the diagonals are 13 cm. Find the area of the trapezoid.

30. We wish to construct a storage box. The top and bottom need to be of a heavier material than the sides. The box measures 27 cm deep, 18 cm wide, and 32 cm long. The material for the sides costs $4.50 a square meter and the material for the top and bottom costs $6.30 a square meter. How much will the material to construct the box cost?

31. A house on a cul-de-sac is advertised as being on a one-third acre lot. The lot is in the shape of an isosceles trapezoid with the front measuring 80 ft., the back measuring 200 ft., and the sides measuring 150 ft. Is it $\frac{1}{3}$ of an acre? (An acre is 43,560 sq. ft.)

*32. In the figure, the side of the large square is 23, the side of the small square is 18. The vertex of the small square is at the center of the large square. A side of the small square intersects a side of the large square 4 units from a vertex. Find the overlapping area.

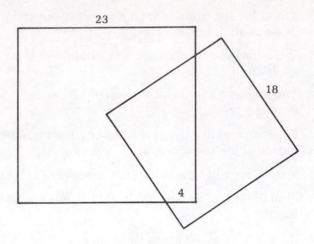

*33. *Viviani's Theorem* (Italian, 1622–1703)

For *any* point $P$ inside an equilateral triangle $ABC$, the sum of the perpendiculars $a$, $b$, and $c$ from $P$ to the sides is equal to the altitude $h$. Prove this theorem. (It is very simple if the correct auxiliary lines are used.)

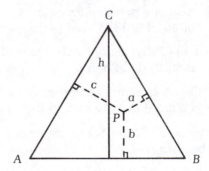

*34. Given: Parallelogram $ABCD$ with $P$ on diagonal $BD$
    $EF \parallel AB$
    $GH \parallel BC$

Prove: $A_{AGPE} = A_{CHPF}$

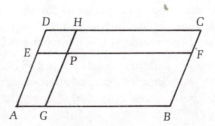

*35. Given *any* triangle with sides $a$, $b$, $c$, when is it possible for the perimeter to be numerically equal to the area?

**\*\*36.** Have you ever noticed that the normal method of cutting a square cake that is frosted on the sides is terribly unfair? Some people get a piece like *A*, with two edges frosted, others get a piece like *B*, with one edge frosted, and some unfortunate souls get a piece like *C*, with no frosted edges.

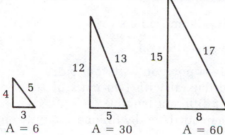

Can you devise a method of cutting a square cake into *any* required number of pieces, so that each piece has the same amount of cake, and also the same amount of frosting? Assume that the top and all edges are frosted.

**\*\*37.** The area of any Pythagorean Triangle (that is, one whose sides are the integers of a Pythagorean Triple) is divisible by 6. Examples:

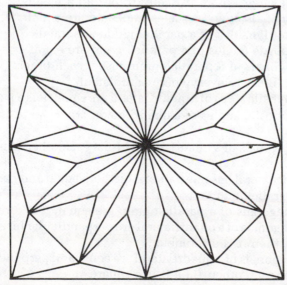

Can you explain why this is so?

**\*38.** A paper published by J. Kurschak in 1898 presented this beautiful pattern. It is called the Kurschak Tile. Use it to prove the theorem (rarely found in any geometry book): *The area of a regular dodecagon is 3r². (r is the radius)* (Hint: prove all the equilateral triangles congruent, prove all the isosceles triangles congruent, move the 6 isosceles and 3 equilateral triangles from inside the dodecahedron in one quadrant, to fill in the squares in the other three quadrants.)\*

\*For more information on the Kurschak Tile see an article by Alexanderson and Seydel in *The Mathematical Gazette;* 1978, Vol. *62,* pp. 192–196.

Theorems about polygons:

The sum of the interior angles of a polygon of $n$ sides is $180(n - 2)$.

In a regular polygon of $n$ sides each interior angle is $\dfrac{180(n - 2)}{n}$.

An exterior angle of a regular polygon of $n$ sides is $\dfrac{360}{n}$.

The sum of the exterior angles of a polygon is 360.

Polygon constructions:

possible:
3-6-12-etc.
4-8-16-etc.
5-10-20-etc.
15-30-60-etc.
$2^m(2^n + 1)$ where $m$ and $n$ are integers and $(2^n + 1)$ is a prime
not possible:
7, 9, 11, 13, and others

Definitions of the quadrilaterals:

Quadrilateral—a polygon with four sides
Kite—a quadrilateral with two pairs of equal consecutive sides, each pair of different length
Trapezoid—a quadrilateral with one and only one pair of opposite sides parallel
Parallelogram—a quadrilateral with its opposite sides parallel
Rectangle—a parallelogram with a right angle
Rhombus—a parallelogram with two adjacent sides equal
Square—a rectangle with two adjacent sides equal, or, a rhombus with a right angle

Theorems about quadrilaterals:

The diagonals of a kite are perpendicular to each other.
The base angles of an isosceles trapezoid are equal.
The diagonals of an isosceles trapezoid are equal.
The opposite sides of a parallelogram are equal.
The opposite angles of a parallelogram are equal.
The diagonals of a parallelogram bisect each other.
Parallel lines are everywhere equidistant.
If the opposite sides of a quadrilateral are equal, it is a parallelogram.
If the opposite angles of a quadrilateral are equal, it is a parallelogram.
If the diagonals of a quadrilateral bisect each other, it is a parallelogram.
If a quadrilateral has one pair of sides both equal and parallel, then it is a parallelogram.
If the diagonals of a parallelogram are equal, it is a rectangle.
If the diagonals of a quadrilateral are perpendicular bisectors of each other, then it is a rhombus.
If the diagonals of a quadrilateral are equal and perpendicular bisectors of each other, then it is a square.

Theorems about areas of plane figures:

The area of a rectangular region is its length times its width. (Def.)

The area of a square is a side squared.

The area of a parallelogram is its base times its height.

The area of a triangle is $\frac{1}{2}$ its base times its height.

Triangles with equal bases and equal heights are equal.

The median of a triangle divides it into two equal triangles.

In a right triangle, the product of the legs equals the product of the hypotenuse and the altitude to the hypotenuse.

The area of an equilateral triangle with side $s$ is $\dfrac{s^2 \sqrt{3}}{4}$.

The area of a triangle is one-half its perimeter times the radius of its inscribed circle. $A = \frac{1}{2}pr$

The area of a rhombus is one-half the product of the diagonals. $A = \frac{1}{2}d_1 d_2$

The area of any triangle is $\sqrt{s(s-a)(s-b)(s-c)}$ where $s$ is the semiperimeter. (Heron's Semiperimeter Formula)

The area of a trapezoid is one-half the product of its height and the sum of its bases. $A = \frac{1}{2}h(b_1 + b_2)$

The area of a regular polygon is one-half its apothem times its perimeter. $A = \frac{1}{2}ap$

*Photograph by Harold M. Lambert.*

# CHAPTER 6

# CIRCLES

Here we investigate the circle and all of its parts. There are a surprising number of theorems that can be written about such a plain and simple thing as the circle. We will also develop many useful formulas about the measure of the parts of a circle and various areas related to the circle.

Two of the supplementary sections can be inserted after the material on circles: Composite figures—in which we find the areas of figures composed of several different parts—and another look at the Pythagorean Theorem, this time considering it as an area relationship, rather than as a numerical one.

Of the myriad shapes that surround us, the circle is one of the most common. Not only is it a shape that is pleasing to the eye, but it is a most efficient shape. It is the shape that encloses the most area with a given amount of perimeter. With a given amount of fencing, more pasture can be enclosed with a circle than with any other configuration. In nature, such forces as gravity, wind resistance, and friction shape heavenly bodies, raindrops, boulders, and even grains of sand into shapes approaching a sphere (the three-dimensional circle).

Circles are the limiting shape of a polygon as the number of sides increases. All properties of polygons which do not depend on the number of sides will also be true for the circle. This is Postulate 7—the Limit Postulate.

A proper definition of a circle needs to be carefully stated. The intuitive notion of a circle as a "round thing" leaves much to be desired.

**21. Circle theorems**

A **circle** is a closed curve in a plane, all points of which are equidistant from a point called the **center.**

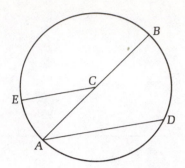

In the figure, *C* is the **center** of the circle.

*CA* and *CB* are **radii** (plural of **radius**), line segments from the center to the circle. Since, by definition, all the points of a cricle are equidistant from the center, **all radii in the same or equal circles are equal.** This corollary to the definition of the circle will be used frequently as a reason in our proofs.

*AB* is a **diameter,** a line segment through the center, terminated at both ends by the circle.

The **circumference** is the measure of the length of the curve.

*AD* is a **chord** of the circle, a line segment connecting two points of the circle.

$\overset{\frown}{AD}$ is an **arc** of the circle, a portion of the circumference. Note the curved line above $\overset{\frown}{AD}$. This prevents confusing arc $\overset{\frown}{AD}$ with chord *AD*.

An arc cut off by a diameter is a **semicircle.**

An arc greater than a semicircle is called a **major arc,** an arc less than a semicircle is called a **minor arc.** A major arc should be named with three letters to prevent confusing it with the minor arc. In the figure $\overset{\frown}{BDE}$ is a major arc.

∢*ACE* is a **central angle,** an angle with its vertex at the center, whose sides are radii.

∢*BAD* is an **inscribed angle,** an angle with its vertex on the circumference of the circle and chords or secants for its rays.

Line (not line segment) *AB* is a **secant,** a line which cuts through the circle at two points.

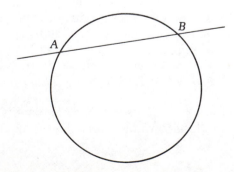

Line *AB* is a **tangent,** a line which intersects the circle at one and only one point.

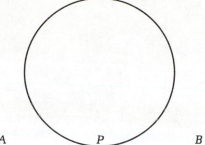

**Concentric** circles are two or more circles which have the same center.

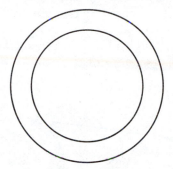

The words **inscribed** and **circumscribed** are adjectives describing figures drawn inside or around other figures. In the figure below, we have a circumscribed quadrilateral and an inscribed circle. Each side of a circumscribed polygon must be a tangent to the circle.

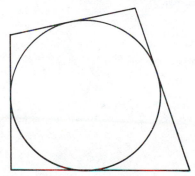

In this figure, we have a circumscribed circle and an inscribed pentagon. No confusion should exist when naming circumscribed and inscribed figures if we first check to see if we are describing the circle or the polygon.

The **line of centers** of two circles is the line segment joining their centers. *AB* is the line of centers in these figures. The circles may be overlapping or separate.

Two or more circles may share the same line as a tangent. When this happens the line is called a **common tangent**. In the figure, *AB* is a **common internal tangent** and *CD* is a **common external tangent**. A common internal tangent cuts through a line of centers if it is drawn, whereas the common external tangent does not.

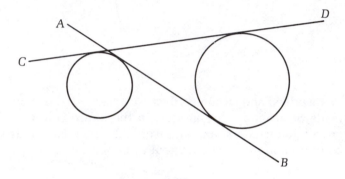

There are several possibilities as to how many common tangents of each type two circles can share.

| # of common internal tangents | # of common external tangents | figure |
|:---:|:---:|:---:|
| 0 | 0 | |
| 0 | 1 | |
| 0 | 2 | |
| 1 | 2 | |
| 2 | 2 | |

Two circles which share a common tangent at a common point are either
**tangent internally** at point P:

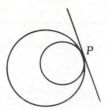

or **tangent externally** at point P:

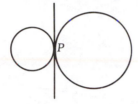

Notice that we are describing the *circles*, and not the tangent.

We should now have sufficient vocabulary to allow us to consider some theorems concerning circles.

The drive belt on the Porsche Turbo-Carrera engine forms external tangents with the drive shaft and the alternator.

> ### *Any three noncollinear points determine a circle.*

*Given:* Points *A*, *B*, and *C*, not on the same line

*Prove:* Circle *O* passes through *A*, *B*, and *C*

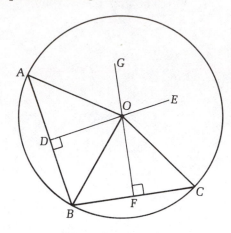

| | |
|---|---|
| **1.** Draw *AB*, *BC* | **1.** Construction (2 pts. determine a line) |
| **2.** Construct *DE* ⊥ bisector of *AB* | **2.** Construction |
| **3.** Construct *FG* ⊥ bisector of *BC* | **3.** Construction |
| **4.** *DE* and *FG* will intersect at some point, *O*. | **4.** 2 lines either intersect or are ‖ . They could be ‖ only if *A*, *B*, and *C* were collinear |
| **5.** Draw *AO*, *BO*, *CO* | **5.** Construction |
| **6.** *BO* is a reflection of *AO* about *DE* | **6.** Def. of reflection |
| **7.** *CO* is a reflection of *BO* about *FG* | **7.** Def. of reflection |
| **8.** *AO* = *BO* = *CO* | **8.** Preservation of lengths |
| **9.** Draw circle *O* with radius *AO* | **9.** Def. of circle |

This theorem shows us how to construct a circle when only an arc, or portion of the circle, is known. Draw two chords and construct their perpendicular bisectors. They will intersect at the center of the circle.

When a pair of glasses are broken, the optician can often determine the prescription from a broken fragment. This method can also be used to find the unique point that is equidistant from three given points. Three houses are to be supplied with power from a single pole. Where should the pole be located so that the distances to all three houses are the same? We can locate the pole, $P$, by constructing the perpendicular bisectors of the lines joining the houses.

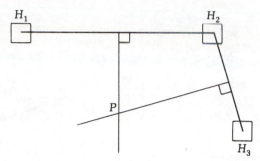

We now wish to prove some relations between three different parts of a circle: a central angle, its arc, and its chord. The theorems apply to parts "in the same or in equal circles." We will illustrate for the case of equal circles. If the same circle is used, the figure needs only to be rotated. **NOTE**—when we say a pair of circles are equal, it can also be understood that they are $\cong$, as all circles have the same shape.

---

*In the same circle or in equal circles, equal central angles have equal arcs.*

---

Given: Circle $C$ = Circle $C'$
    $\angle ACB = \angle A'C'B'$

Prove: $\overarc{AB} = \overarc{A'B'}$

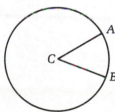    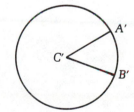

1. Circle $C$ = Circle $C'$                1. Given

2. Translate $C'$ to $C$ so that           2. Existence of a translation
   the circles coincide and
   rotate so that $C'A'$
   falls on $CA$

3. $\angle ACB = \angle A'C'B'$            3. Given

4. $CB$ falls on $C'B'$                     4. Angle Measure Postulate

5. $\overarc{AB} = \overarc{A'B'}$         5. They coincide

> *In the same circle or in equal circles, equal arcs have equal central angles.*

Given:  Circle $C$ = Circle $C'$
$\quad\quad\quad\widehat{AB} = \widehat{A'B'}$

Prove:  $\angle ACB = \angle A'C'B'$

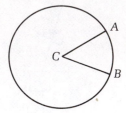 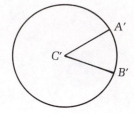

| | |
|---|---|
| **1.** Circle $C$ = Circle $C'$ | **1.** Given |
| **2.** Translate $C'$ to $C$ so that the circles coincide and rotate so that $C'A'$ falls on $CA$ | **2.** Existence of a translation |
| **3.** $\widehat{AB} = \widehat{A'B'}$ | **3.** Given |
| **4.** $CA = CB = C'A' = C'B'$ | **4.** Radii of the same circle |
| **5.** $\angle ACB = \angle A'C'B'$ | **5.** They coincide |

**NOTE** that we could also show this by appealing to the definition of an angle being an amount of rotation: equal rotations (arcs) would therefore give equal angles.

> *In the same circle or in equal circles, equal arcs have equal chords.*

Given:  Circle $C$ = Circle $C'$
$\quad\quad\quad \overset{\frown}{AB} = \overset{\frown}{A'B'}$

Prove:  $AB = A'B'$

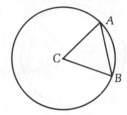 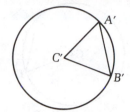

| | |
|---|---|
| **1.** Circle $C$ = Circle $C'$ | **1.** Given |
| **2.** Draw $AB$ and $A'B'$ | **2.** Construction (2 pts. determine a line) |
| **3.** $\overset{\frown}{AB} = \overset{\frown}{A'B'}$ | **3.** Given |
| **4.** $\sphericalangle ACB = \sphericalangle A'C'B'$ | **4.** = arcs have = central angles |
| **5.** $AC = BC = A'C' = B'C'$ | **5.** Radii of equal circles |
| **6.** $\triangle ACB \cong \triangle A'C'B'$ | **6.** SAS |
| **7.** $AB = A'B'$ | **7.** CPCTE |

Circular design in stone.

> *In the same circle or in equal circles, equal chords have equal arcs.*

Given: Circle $C$ = Circle $C'$
       $AB = A'B'$

Prove: $\widehat{AB} = \widehat{A'B'}$

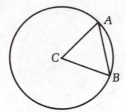

| | |
|---|---|
| **1.** Circle $C$ = Circle $C'$ | **1.** Given |
| **2.** $CA = CB = C'A' = C'B'$ | **2.** Radii of same or equal circles |
| **3.** $AB = A'B'$ | **3.** Given |
| **4.** $\triangle ACB \cong \triangle A'C'B'$ | **4.** SSS |
| **5.** $\angle ACB = \angle A'C'B'$ | **5.** CPCTE |
| **6.** $\widehat{AB} = \widehat{A'B'}$ | **6.** = central angles have = arcs |

> *In the same circle or in equal circles, equal central angles have equal chords.*

Given: Circle $C$ = Circle $C'$
       $\angle ACB = \angle A'C'B'$

Prove: $AB = A'B'$

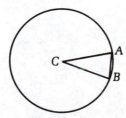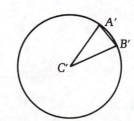

| | |
|---|---|
| **1.** Circle $C$ = Circle $C'$ | **1.** Given |
| **2.** $\angle ACB = \angle A'C'B'$ | **2.** Given |
| **3.** $\widehat{AB} = \widehat{A'B'}$ | **3.** = central angles have = arcs |
| **4.** $AB = A'B'$ | **4.** = arcs have = chords |

> *In the same circle or in equal circles, equal chords have equal central angles.*

Given:  Circle $C$ = Circle $C'$
     $AB = A'B'$

Prove:  $\angle ACB = \angle A'C'B'$

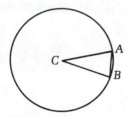 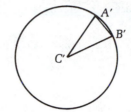

1. Circle $C$ = Circle $C'$     1. Given

2. $AB = A'B'$     2. Given

3. $\overset{\frown}{AB} = \overset{\frown}{A'B'}$     3. = chords have = arcs

4. $\angle ACB = \angle A'C'B'$     4. = arcs have = central angles

To summarize these six theorems, we have the following relationships:

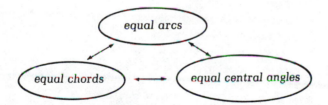

The following theorem, while not easy to prove, presents a relationship you were probably intuitively aware of. It will be useful in numerical relationships and in finding the areas of polygons.

> *A line through the center of a circle, perpendicular to a chord, bisects the chord and its arcs.*

Given: DE passes through $O$
$OC \perp AB$

Prove: $AC = BC$, $\widehat{AD} = \widehat{BD}$, $\widehat{AE} = \widehat{BE}$

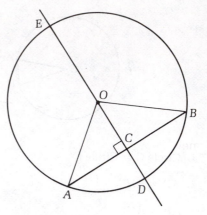

| 1. $OC \perp AB$ | 1. Given |
|---|---|
| 2. $\angle OCA = \angle OCB$ | 2. Rt. angles are = |
| 3. Draw $OA$ and $OB$ | 3. Construction (2 pts. determine a line) |
| 4. $OA = OB$ | 4. Radii of same circle are = |
| 5. $\angle A = \angle B$ | 5. Angles opp. = sides are = |
| 6. $\angle AOC = \angle BOC$ | 6. If 2 angles of a $\triangle$ = 2 angles of a 2nd $\triangle$, the third angles are = |
| 7. $\widehat{AD} = \widehat{BD}$ | 7. = central angles have = arcs |
| 8. $OC = OC$ | 8. Reflexive |
| 9. $\triangle AOC \cong \triangle BOC$ | 9. ASA |
| 10. $AC = BC$ | 10. CPCTE |
| 11. $\angle AOC$ and $\angle AOE$ are supplements $\angle BOC$ and $\angle BOE$ are supplements | 11. Def. of supp. |
| 12. $\angle AOE = \angle BOE$ | 12. Supp. of equals are equal |
| 13. $\widehat{AE} = \widehat{BE}$ | 13. = central angles have = arcs |

Notice the plural form, "arcs," in this theorem. Both the major arc of the chord and the minor arc of the chord are bisected.

When the hypothesis and the conclusion of a conditional statement are interchanged, the resulting statement is called the converse. When a statement has several parts in its hypothesis or conclusion, and some but not all of them are interchanged, the resulting statement is called a partial converse.

The preceding theorem has the following partial converse:

> *The perpendicular bisector of a chord passes through the center of the circle.*

Since we have shown that the center of a circle can be found by constructing the perpendicular bisectors of two chords, the center must lie somewhere on any one perpendicular bisector.

For the following theorem and its converse, we will illustrate the case for "same circle." If the situation is "equal circles" simply translate so that the centers coincide.

> *In the same circle or in equal circles, equal chords are*
> *equidistant from the center.*

Given: $AB = CD$

$OX \perp AB, OY \perp CD$ (Distance from a point to a line is the shortest distance, which is measured on a perpendicular.)

Prove: $OX = OY$

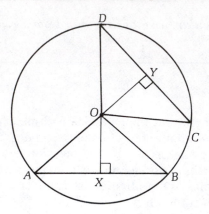

(Analysis: We will rotate one triangle so that we can show OX and OY coincide.)

| | |
|---|---|
| 1. Draw $OA$, $OB$, $OC$, $OD$ | 1. Construction |
| 2. Rotate $\triangle COD$ so that $CO$ coincides with $AO$ | 2. Existence of a rotation |
| 3. $AB = CD$ | 3. Given |
| 4. $\measuredangle AOB = \measuredangle COD$ | 4. = chords have = cent. angles |
| 5. $OD$ will fall on $OB$ | 5. Angle Measure Post. |
| 6. $B$ and $D$ will coincide | 6. Distance Postulate |
| 7. $OX \perp AB$ and $OY \perp CD$ | 7. Given |
| 8. $X$ and $Y$ are midpoints | 8. Line $\perp$ to a chord and through the center of a circle bis. the chord |
| 9. $X$ and $Y$ coincide | 9. Halves of equals are equal |
| 10. $OX = OY$ | 10. 2 pts. determine a unique line |

> *In the same circle or in equal circles, chords equidistant from the center are equal.*

Given:  $OX = OY$
  $OX \perp AB$, $OY \perp CD$

Prove:  $AB = CD$

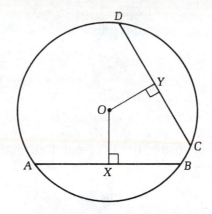

| | |
|---|---|
| **1.** Rotate $OY$ so as to fall on $OX$ | **1.** Existence of a rotation |
| **2.** $OX = OY$ | **2.** Given |
| **3.** $X$ falls on $Y$ | **3.** Distance Post. |
| **4.** $OX \perp AB$ and $OY \perp CD$ | **4.** Given |
| **5.** $AB$ coincides with $CD$ | **5.** Construction (only 1 $\perp$ to a line at a point) |
| **6.** $AB = CD$ | **6.** They coincide |

In order to construct a tangent to a circle at a given point $P$ on the circle, first draw the radius $OP$ extended, then construct the perpendicular to $OP$ at $P$. This will be a tangent.

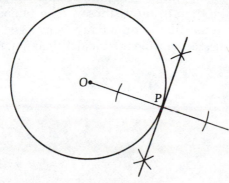

The next theorem confirms this relationship.

> ### *The perpendicular to a radius at its extremity is a tangent.*

Given:  Radius $OX$ of circle $O$
$\qquad OX \perp AB$

Prove:  $AB$ is a tangent

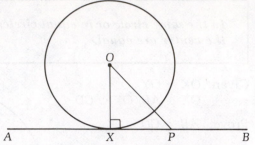

(Analysis: In order to show that $AB$ is a tangent, we need to use the definition. That is, show that $X$ is the only point of the line that lies on the circle. We can do this by showing *any* (every) other point of $AB$ is a distance greater than $OX$ from the center, and thus is not on the circle.)

| | |
|---|---|
| **1.** Choose *any* other point $P$ on $AB$ and draw $OP$ | **1.** Construction (2 pts. determine a line) |
| **2.** $OX \perp AB$ | **2.** Given |
| **3.** $OP$ is hypotenuse of rt. $\triangle OPX$ | **3.** Def. of hypotenuse |
| **4.** $OP > OX$ | **4.** Pythagorean Relation (with some algebra) |
| **5.** $P$ lies outside circle $O$ | **5.** $OP$ greater than a radius |
| **6.** $AB$ is a tangent | **6.** Only one point of intersection with circle $O$ |

This theorem has several partial converses that are also true, two of which are: "*A perpendicular to a tangent at its point of contact passes through the center of the circle;*" and "*A perpendicular to a tangent, through the center of a circle, will pass through the point of contact.*" We will not take the time to prove these, since understanding this basic relationship will usually be sufficient for our needs. One of the partial converses is necessary for the next major theorem, however, so we will prove it now.

> *A radius drawn to the point of contact of a tangent is perpendicular to the tangent.*

Given:  Tangent *AB* intersecting circle *O* at *P*

Prove:  *OP* ⊥ *AB*

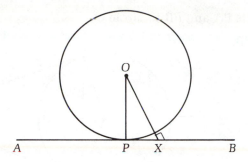

(Analysis: We will show *OP* ⊥ *AB* by an indirect proof, method of reductio ad absurdum.)

Assume *OP* is not perpendicular to *AB*  |  Assumption

1. Draw *OX* ⊥ *AB*  |  1. Construction (only 1 ⊥ from a pt. to a line)

2. *AB* is a tangent  |  2. Given

3. *OX* > *OP*  |  3. *X* is outside the circle *P* is on the circle

4. *OP* > *OX*  |  4. Hypotenuse is > a leg of rt. triangle *OPX*

5. ∴ Contradiction

6. ∴ *OP* ⊥ *AB*

We can now prove deductively this relationship that seems intuitively "obvious":

---

### Tangents to a circle from an external point are equal.

*Given:* Tangents *PA* and *PB* to circle *O*

*Prove:* *PA* = *PB*

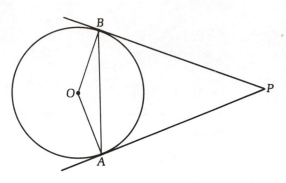

| 1. *PA* and *PB* are tangents | 1. Given |
| 2. Draw *OA* and *OB* | 2. Construction |
| 3. *OA* ⊥ *PA* and *OB* ⊥ *PB* | 3. Radius drawn to a tangent is ⊥ |
| 4. ∡*OAB* and ∡*PAB* are complementary and ∡*OBA* and ∡*PBA* are complementary | 4. They form a rt. angle |
| 5. *OA* = *OB* | 5. Radii of the same circle |
| 6. ∡*OAB* = ∡*OBA* | 6. Angles opp. = sides are = |
| 7. ∡*PAB* = ∡*PBA* | 7. Complements of equals are equal |
| 8. *PA* = *PB* | 8. Sides opp. = angles are = |

---

Tangents to a circle have many applications for us on the sphere which we inhabit. We cannot see a ship beyond the horizon unless we climb a tall mast.

(not to scale)

We cannot send direct television signals to a distant city unless we have a tall antenna.

With a given height, what distance can be seen? Or conversely, what height is necessary to see a given distance? We can calculate the relationship between distance and height with a theorem from a previous chapter: The altitude to the hypotenuse of a right triangle is the mean proportional between the segments of the hypotenuse. Looking at a cross-section of the earth we see that the tangent $AC$ is perpendicular to the radius $OA$.

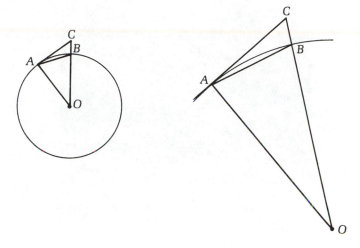

Now, for reasonable distances (less than 1200 km), ∢$ABO$ is nearly a right angle, and $\overparen{AB}$ is nearly equal to $AB$. So we get good approximations if we consider $AB$ to be the altitude to the hypotenuse of a right triangle.

$$AB = \text{the distance, } d$$
$$BC = \text{the height, } h$$
$$OB = \text{the radius of the earth, } r; \text{ therefore}$$

$$\frac{h}{d} = \frac{d}{r} \text{ or } d = \sqrt{hr} \text{ where } r \approx 6370 \text{ km or } 3960 \text{ mi.}$$

In this type of application we are likely to have mixed denominations: $r$ in kilometers or miles, $h$ in meters or feet, and $d$ in kilometers or miles. So we get:

$$d \text{ (in km)} = \sqrt{\frac{h(\text{in m})}{1000} \times r(\text{in km})} = \sqrt{\frac{6370}{1000}h} \approx 2.52 \sqrt{h}$$

or

$$d \text{ (in mi.)} = \sqrt{\frac{h(\text{in ft.})}{5280} \times r(\text{in mi.})} = \sqrt{\frac{3960}{5280}h} \approx .87 \sqrt{h}$$

These formulas are then easy to use to solve our visible distance problems.

Sutro tower, near San Francisco, is 580 ft. tall, and is built on a 910 ft. high hill. By the formula on page 259, it could send direct signals a distance of 34 miles.

We now prove one final relationship regarding parts of a circle.

> *Two parallel lines intercept equal arcs on a circle.*

This theorem needs to be considered in three parts, since the parallel lines can be:

| a tangent and a secant | 2 secants | or 2 tangents |
|:---:|:---:|:---:|
| 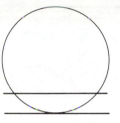 | 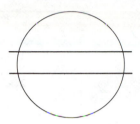 |  |
| Part I | Part II | Part III |

## Part I

Given: $AB \parallel CD$
$\quad\quad$ $AB$ is a tangent at $E$

Prove: $\overset{\frown}{CE} = \overset{\frown}{DE}$

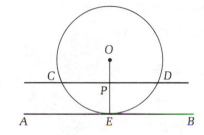

| | |
|---|---|
| **1.** Draw $OE$ | **1.** Construction (2 pts. determine a line) |
| **2.** $AB$ is tangent at $E$ | **2.** Given |
| **3.** $OE \perp AB$ | **3.** Radius drawn to tangent is $\perp$ |
| **4.** $\angle OEB$ is a rt. $\angle$ | **4.** $\perp$ s form rt. $\angle$ s |
| **5.** $CD \parallel AB$ | **5.** Given |
| **6.** $\angle OPD$ is a rt. $\angle$ | **6.** Corresponding angles |
| **7.** $OE \perp CD$ | **7.** Def. of $\perp$ |
| **8.** $\overset{\frown}{CE} = \overset{\frown}{DE}$ | **8.** Line $\perp$ to a chord bisects the chord and its arcs |

*Proof continued on the following page*

**Part II**

Given: $AB \parallel CD$

Prove: $\overset{\frown}{AC} = \overset{\frown}{BD}$

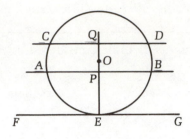

| | |
|---|---|
| **1.** Draw $OE \perp AB$ | **1.** Construction |
| **2.** $\sphericalangle OPB$ is a rt. $\sphericalangle$ | **2.** $\perp$s form rt. $\sphericalangle$s |
| **3.** $AB \parallel CD$ | **3.** Given |
| **4.** $\sphericalangle OQC$ is a rt. $\sphericalangle$ | **4.** Alt. int. angles |
| **5.** $OE \perp CD$ | **5.** Def. of $\perp$ |
| **6.** Construct $FG \perp OE$ at $E$ | **6.** Construction |
| **7.** $FG \parallel CD \parallel AB$ | **7.** Lines $\parallel$ to same line are $\parallel$ |
| **8.** $\overset{\frown}{CE} = \overset{\frown}{DE}$ | **8.** Part I |
| **9.** $\overset{\frown}{AE} = \overset{\frown}{BE}$ | **9.** Part I |
| **10.** $\overset{\frown}{AC} = \overset{\frown}{BD}$ | **10.** Subtraction of (9) from (8) |

**Part III**

Given: $AB \parallel CD$

Prove: $\overset{\frown}{EGF} = \overset{\frown}{EHF}$

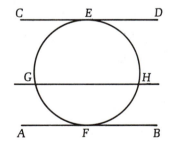

| | |
|---|---|
| **1.** Draw $GH \parallel AB$ | **1.** Construction |
| **2.** $CD \parallel AB$ | **2.** Given |
| **3.** $CD \parallel GH$ | **3.** Lines $\parallel$ to same line are $\parallel$ |
| **4.** $\overset{\frown}{GE} = \overset{\frown}{HE}$ | **4.** Part I |
| **5.** $\overset{\frown}{GF} = \overset{\frown}{HF}$ | **5.** Part I |
| **6.** $\overset{\frown}{EGF} = \overset{\frown}{EHF}$ | **6.** Addition |

We have presented quite a few circle relationships in this section. Perhaps we had better summarize.

*The six relationships with central angles, arcs, and chords:*

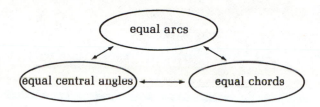

*Any three noncollinear points determine a circle.*
*A radius perpendicular to a chord bisects the chord and its arcs.*
*The perpendicular bisector of a chord passes through the center of the circle.*
*Equal chords are equidistant from the center.*
*Chords equidistant from the center are equal.*
*A perpendicular to a radius at its extremity is a tangent.*
*A radius drawn to the point of contact of a tangent is perpendicular to the tangent.*
*Tangents to a circle from an external point are equal.*
*Parallel lines intercept equal arcs on a circle.*

Of course part of the hypothesis for many of the above theorems was "in the same circle or in equal circles".

Let us finish this section with some examples of how these theorems can be used in proofs and in numerical situations.

**Given:** $AB = CD$

**Prove:** $AC = BD$

1. $AB = CD$
2. $\overarc{AB} = \overarc{CD}$
3. $\overarc{CB} = \overarc{CB}$
4. $\overarc{AC} = \overarc{BD}$
5. $AC = BD$

1. Given
2. = chords have = arcs
3. Reflexive
4. Subtraction
5. = arcs have = chords

**Given:** Concentric circles, with chord $AD$ of the larger circle cutting the smaller circle at $B$ and $C$

**Prove:** $AB = CD$

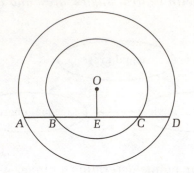

1. Draw $OE \perp AD$                1. Construction

2. $AE = DE$                          2. Radius $\perp$ to a chord bis. it

3. $BE = CE$                          3. Radius $\perp$ to a chord bis. it

4. $AB = CD$                          4. Subtraction

**Two tangents, $PB$ and $PA$, are drawn to circle $O$ from point $P$. $\measuredangle AOB = 120°$ and $OA = 7$ cm. Find the perimeter of quadrilateral $AOBP$.**

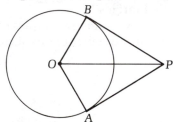

Since radii to a tangent are perpendicular, radii of the same circle are equal, and tangents to a circle from an external point are equal, we can determine that the triangles are congruent and that each is a 30-60-90 triangle. $OA = OB = 7$, $PA = PB = 7\sqrt{3}$. So the perimeter of $AOBP = 14 + 14\sqrt{3}$ cm.

**Triangle $ABC$ with sides 12, 16, 20 is circumscribed about a circle with points of tangency $P$, $Q$, and $R$. Find $AP$, $PB$, $BQ$, $QC$, $CR$, and $RA$.**

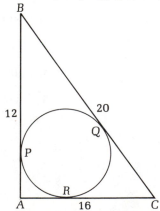

Call one segment $X$ and solve algebraically, using equal tangents.

**265**

**21–CIRCLE THEOREMS**

$$AP = X, PB = 12 - X, BQ = 12 - X$$
also $RA = X$, $RC = 16 - X$, $CQ = 16 - X$
so on $CB$, $16 - X + 12 - X = 20$, $X = 4$
therefore $AP = 4$, $PB = 8$, $BQ = 8$, $QC = 12$, $CR = 12$, $RA = 4$

**Find the perimeter of trapezoid $ABCD$, tangent to a circle at $P$, $Q$, $R$, $S$ with bases of 5 cm and 12 cm.**

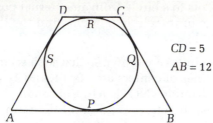

$CD = 5$
$AB = 12$

$$QC = RC \text{ and } DS = DR$$
so since $DR + RC = 5$, $DS + CQ = 5$
likewise $AS + BQ = 12$.
Therefore, the perimeter of $ABCD = 5 + 5 + 12 + 12 = 34$ cm

1. $COD$ is a diameter of circle $O$, radius $OF$ is parallel to chord $CE$.
   Prove that $\overarc{EF} = \overarc{FD}$
   Hint: Draw $EO$

   **EXERCISE SET 21**

   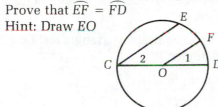

2. $AB = CD$. Prove $AC = DB$
   Hint: $\overarc{AB} = \overarc{CD}$, what arc is common?

   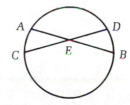

3. Given two concentric circles with chord $AD$ of the larger circle cutting the smaller circle at $B$ and $C$. Prove that $AB = CD$

   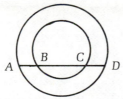

4. Given that $AOD$ is a diameter of circle $O$, chord $AB$ = chord $CD$. Prove that $AB \parallel CD$

5. Tangents to a circle from an external point are each 9 cm long and form an angle of 60° at that point. How long is the chord joining their points of contact?

6. An isosceles trapezoid is circumscribed about a circle. The bases are 14 cm and 8 cm. Find the perimeter of the trapezoid.

7. A circle is inscribed in a right triangle whose sides are 3, 4, and 5 cm. Find the radius of the circle.

8. Tangents to a circle from an external point $P$ form an angle of 60°. If the radius of the circle is 5 cm, find the distance from $P$ to the center of the circle.

9. Three circles are drawn so that each is tangent externally to the other two. The lines of centers are 10, 14, and 18 cm respectively. Find the radius of each circle. (NOTE: a line of center passes through the point of contact and is, therefore, the sum of the radii of two circles.)

10. It is always possible to draw a circle through three noncollinear points. When is it possible to draw a circle through four points which are noncollinear?

11. If a sailor climbs to the top of a 10 m mast, how far can he see? (That is, how far is it to his horizon?)

12. The tallest man-made structure in the U.S. is the television tower of station KTHI in North Dakota. It is 2063 ft. tall. What is its line-of-sight broadcasting range?

13. What is the side of an equilateral triangle just large enough to circumscribe a circle of diameter 1?

14. *Given:* M is the midpoint of AB

*Prove:* $\angle AOM = \angle BOM$

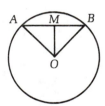

15. *Given:* $\angle AOB = \angle COD$

*Prove:* $AC = BD$

16. Draw three circles situated so that they have no points of tangency. Rearrange them so that there is only one point of tangency, then 2 and 3 points of tangency.

17. Can you draw three circles with an external tangent common to all three?

18. Can you draw three circles with an internal tangent common to all three? (It would have to cut through all three lines of centers.)

19. Given: $OC \perp AB$

   Prove: $\angle 1 = \angle 2$

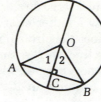
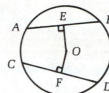

20. Given: $AB = CD$
   $OE \perp AB$
   $OF \perp CD$

   Prove: $OE = OF$

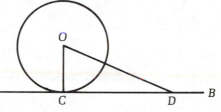

21. Given: $OC = 5$
   $CD = 12$
   $OD = 13$

   Prove: $AB$ is a tangent

22. Given: $PA$, $PB$, and $PC$ are tangents

   Prove: $PA = PC$

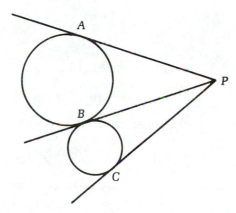

23. Tangents to a circle from an external point form an angle of 120°. The line from the external point to the center of the circle is 13 cm. Find the perimeter of the figure formed by the two tangents and the radii to the points of contact.

24. Triangle $PRQ$ is circumscribed about a circle. $PQ = 10$, $QR = 26$, $PR = 24$. Find the radius of the circle.

25. An isosceles trapezoid is circumscribed about a circle. Its upper base is 15 cm and its lower base is 23 cm. Find its perimeter.

*26. Given: Isosceles trapezoid $ABCD$ circumscribed about circle $O$

   Prove: $DS = CS$

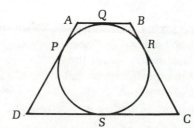

*27. Draw a large circle. Divide the circle into two parts by drawing a chord. Draw another chord, maximizing the number of parts. How many parts does the circle now have? Draw a third chord, not allowing three or more lines to pass through the same point. Into how many parts has the circle now been divided? Continue the process and see if you can generalize a formula for the number of parts in a circle divided by $n$ chords.

 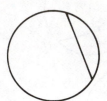  

| # of Chords | # of Parts |
|:-----------:|:----------:|
| 0 | 1 |
| 1 | 2 |
| 2 | 4 |
| 3 | 7 |
| 4 | ? |
| n | $f(n)$ |

*28. Given: radius $OC$
  $P$ on $OC$
  $\angle APC = \angle BPC$

Prove:  $PA = PB$

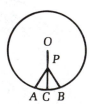

**22. Angle and arc measure**

Both angles and arcs can be measured in terms of degrees. But an angle of 40° is certainly a different sort of thing than an arc of 40°.

Some texts differentiate between the two and would say that $\angle A \overset{m}{=} \overset{\frown}{BC}$, where $\overset{m}{=}$ is read as "has the same measure as" or "is measured by". We will not worry about the distinction between these two types of degrees because both can be considered as the result of a single rotation; and with any given angle, we can always associate an arc with the same measure, and with any given arc, we can associate an angle with the same measure.

We have defined an angle as the result of a rotation of one ray from a position of another. As *AB'* rotates away from *AB*, every point on *AB* will describe an arc.

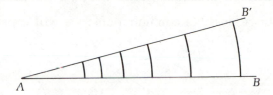

Since one full revolution is 360°, an arc of a full circle has a degree measure of 360°, and any given arc will have the same degree measure as the angle of rotation that generated it.

Don't be confused by the fact that a 40° arc of a circle with radius of 10 cm will be longer than a 40° arc of a circle with radius of 2 cm. They both measure 40° **of arc,** or $\frac{1}{9}$ of the total arc of the circle.

On the basis of this reasoning we will make the following definition of degrees of arc: ***An arc has the same measure as its central angle.***

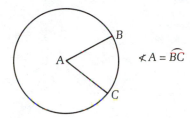

From this beginning, we can now discuss some other relationships between angles and arcs in a circle.

Use of arcs in architectural design. *Stanford Memorial Chapel.*

> *An inscribed angle is half the measure of its intercepted arc.*

Since there are three possible configurations, we will consider three cases:

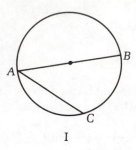 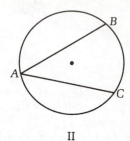 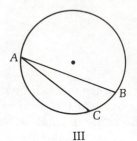

      I                     II                    III

Given: $\angle A$ inscribed in circle $O$.

Prove: $\angle A = \frac{1}{2}\overset{\frown}{BC}$

---

**Part I**

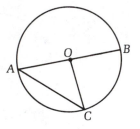

| | |
|---|---|
| **1.** Draw $OC$ | **1.** Construction |
| **2.** $OA = OB$ | **2.** Radii of the same circle are = |
| **3.** $\angle A = \angle C$ | **3.** Angles opp. = sides are = |
| **4.** $\angle BOC = \angle A + \angle C$ | **4.** Exterior angle = sum of opp. interior $\angle$s |
| **5.** $\angle BOC = 2\angle A$ | **5.** Substitution |
| **6.** $\angle A = \frac{1}{2}\angle BOC$ | **6.** Division |
| **7.** $\angle BOC = \overset{\frown}{BC}$ | **7.** Central $\angle$ = intercepted arc |
| **8.** $\angle A = \frac{1}{2}\overset{\frown}{BC}$ | **8.** Substitution |

**Part II**

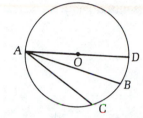

1. Draw diameter $AOD$

2. $\angle 1 = \frac{1}{2}\widehat{BD}$

3. $\angle 2 = \frac{1}{2}\widehat{CD}$

4. $\angle BAC = \frac{1}{2}\widehat{BC}$

1. Construction

2. Part I

3. Part I

4. Addition

**Part III**

1. Draw diameter $AOD$

2. $\angle CAD = \frac{1}{2}\widehat{CD}$

3. $\angle BAD = \frac{1}{2}\widehat{BD}$

4. $\angle BAC = \frac{1}{2}\widehat{BC}$

1. Construction

2. Part I

3. Part I

4. Subtraction

This theorem has some very interesting and useful corollaries.

> *Angles inscribed in the same arc or in equal arcs are equal.*

First the case of angles inscribed in the same arc.

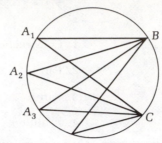

No matter where on the circumference of the circle you locate vertex $A$, $\angle A$ will always have the same measure.

Given: Angles $A_1$, $A_2$, $A_3$, . . . inscribed in $\overset{\frown}{BC}$

Prove: $\angle A_1 = \angle A_2 = \angle A_3 = \ldots$

| | |
|---|---|
| **1.** Angles $A_1$, $A_2$, $A_3$, . . . inscribed in $\overset{\frown}{BC}$ | **1.** Given |
| **2.** $\angle A_1 = \frac{1}{2}\overset{\frown}{BC}$, $\angle A_2 = \frac{1}{2}\overset{\frown}{BC}$, etc. | **2.** Inscribed $\angle$ is $\frac{1}{2}$ arc |
| **3.** $\angle A_1 = \angle A_2 = \angle A_3 = \ldots$ | **3.** Substitution |

For the case of angles inscribed in equal arcs, one more substitution is necessary:

Given: $\overset{\frown}{BC} = \overset{\frown}{EF}$

Prove: $\angle A = \angle D$

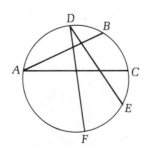

| | |
|---|---|
| **1.** $\angle A = \frac{1}{2}\overset{\frown}{BC}$ | **1.** Inscribed angle |
| **2.** $\angle D = \frac{1}{2}\overset{\frown}{EF}$ | **2.** Inscribed angle |
| **3.** $\overset{\frown}{BC} = \overset{\frown}{EF}$ | **3.** Given |
| **4.** $\angle A = \angle D$ | **4.** Substitution |

Another useful corollary is:

---

*An angle inscribed in a semicircle is a right angle.*

---

Given: Diameter $AOC$
Inscribed angle $ABC$

Prove: $\sphericalangle ABC$ is a right angle

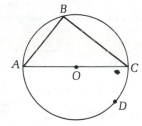

1. $AOC$ is a diameter      1. Given

2. $\overset{\frown}{ADC} = 180°$      2. Def. of diameter

3. $\sphericalangle ABC = \frac{1}{2}(180) = 90°$      3. Inscribed angle

4. $\sphericalangle ABC$ is a right angle      4. Def. of rt. angle

A great many uses can be found for this theorem and its corollaries.

A craftsman making a bowl (or any other object) in the shape of a hemisphere and wishing to test its "roundness" needs only to insert a right angle:

 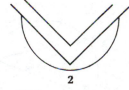 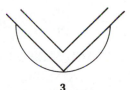

   **1**                    **2**                   **3**

One of three things will happen—either the right angle touches one edge and the bottom, and not the other edge (1); the right angle touches both edges, but not the bottom (2); or the right angle touches both edges and the bottom (3). In the first case, the bowl is not deep enough at that particular point. In the second case, the bowl is too deep at that particular point. In the third case, the bowl is exactly a semicircle in that particular plane. (This sounds like *Goldilocks and the Three Bears!*)

If, as the right angle is moved around the bottom of the bowl, the legs of the angle continue to touch the bowl's rim, then it is a perfect hemisphere.

In the section on the Pythagorean Theorem we discussed a method for constructing any irrational length of the form $\sqrt{N}$. This corollary ("An angle inscribed in a semicircle is a right angle."), along with the theorem that the altitude to the hypotenuse of a right triangle is the mean proportional to the segments of the hypotenuse, gives us an easier method to construct the square root of any counting number.

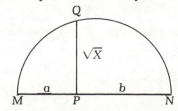

**To construct $\sqrt{X}$:**

**Find any two factors of $X$, such that a $\times$ b = $X$ (If $X$ is prime use $1 \times X$)**

**Mark off $a$ units and then $b$ units on a line**

**Bisect the segment $a + b$, and use this center point to draw a semicircle on $a + b$**

**Construct a perpendicular to $a + b$ at $P$**

**$PQ = \sqrt{X}$**

*Proof:* (if you can supply the reasons)

1. Draw MQ & NQ

2. $\angle MQN$ is a right angle

3. $\triangle MQN$ is a right triangle

4. $\dfrac{a}{PQ} = \dfrac{PQ}{b}$

5. $PQ^2 = a \times b$

6. $PQ^2 = X$

7. $PQ = \sqrt{X}$

Here are a couple of exercises making use of the previous theorems:

**Given:** $AC = CE$

**Prove:** $\triangle ACD \cong \triangle ECB$

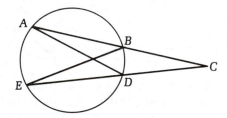

| | |
|---|---|
| 1. $AC = CE$ | 1. Given |
| 2. $\angle C = \angle C$ | 2. Reflexive |
| 3. $\angle A = \angle E$ | 3. $\angle$s inscribed in same arc are $=$ |
| 4. $\triangle ACD \cong \triangle ECB$ | 4. ASA |

**Given:** $\overset{\frown}{AD} = 62°$
$\overset{\frown}{BC} = 128°$

**Find:** ∢$APD$

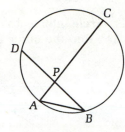

1. ∢$APD$ = ∢$A$ + ∢$B$

2. ∢$A$ = $\frac{1}{2}\overset{\frown}{BC}$ = $\frac{1}{2}(128)$ = 64°

3. ∢$B$ = $\frac{1}{2}\overset{\frown}{AD}$ = $\frac{1}{2}(62)$ = 31°

4. ∢$APD$ = 64 + 31 = 95°

1. Exterior angle = sum of opp. interior ∢s

2. Inscribed angle is $\frac{1}{2}$ intercepted arc

3. (2)

4. Substitution

A method similar to the example above is used to generalize a method for finding the size of angles formed by two intersecting chords of a circle.

---

*An angle formed by two intersecting chords is half the sum of the two intercepted arcs.*

---

**Given:** Chords $AB$ and $CD$ intersecting at $E$

**Prove:** ∢$AEC$ = $\frac{1}{2}(\overset{\frown}{AC} + \overset{\frown}{BD})$

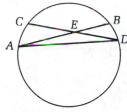

1. Draw $AD$

2. ∢$AEC$ = ∢$A$ + ∢$D$

3. ∢$A$ = $\frac{1}{2}\overset{\frown}{BD}$

4. ∢$D$ = $\frac{1}{2}\overset{\frown}{AC}$

5. ∢$AEC$ = $\frac{1}{2}(\overset{\frown}{AC} + \overset{\frown}{BD})$

1. Construction

2. Exterior angle = sum of opp. interior ∢s

3. Inscribed angle

4. Inscribed angle

5. Substitution

This theorem can be used to find arc measures as well as angles:

**Two chords of a circle intersect with an angle of 42°. One of the intercepted arcs is 54°; find the other arc.**

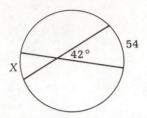

Since $42 = \frac{1}{2}(X + 54)$, $84 = X + 54$, $X = 30$. The other arc measures 30°

Another theorem relating the sizes of angles and arcs states:

---

*An angle formed by a chord and a tangent at one end of the chord is half the intercepted arc.*

---

Given: Chord $AC$, tangent $AB$

Prove: $\angle CAB = \frac{1}{2}\overset{\frown}{AC}$

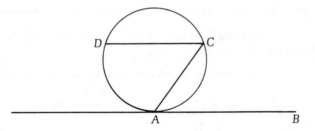

| | |
|---|---|
| **1.** Draw $CD \parallel AB$ | **1.** Construction |
| **2.** $\angle C = \frac{1}{2}\overset{\frown}{AD}$ | **2.** Inscribed angle |
| **3.** $\angle C = \angle CAB$ | **3.** Alt. int. angles |
| **4.** $\overset{\frown}{AD} = \overset{\frown}{AC}$ | **4.** $\parallel$ lines intercept = arcs |
| **5.** $\angle CAB = \frac{1}{2}\overset{\frown}{AC}$ | **5.** Substitution |

Finally, one more theorem which yields the same numerical relationship for three different configurations.

*An angle formed by two secants, by a secant and a tangent, or
by two tangents is half the difference of the intercepted arcs.*

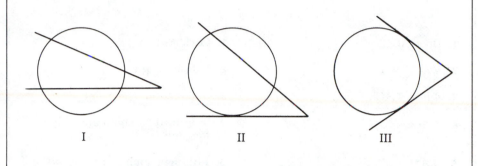

|   |   |   |
|---|---|---|
| I | II | III |

**Part I**

Given: Secants $AB$ and $BC$

Prove: $\angle B = \frac{1}{2}(\widehat{AC} - \widehat{DE})$

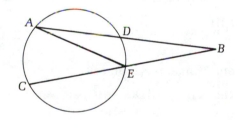

1. Draw $AE$                          1. Construction

2. $\angle AEC = \angle B + \angle A$          2. Exterior angle

3. $\angle B = \angle AEC - \angle A$          3. Subtraction

4. $\angle AEC = \frac{1}{2}\widehat{AC}$          4. Inscribed angle

5. $\angle A = \frac{1}{2}\widehat{DE}$          5. Inscribed angle

6. $\angle B = \frac{1}{2}(\widehat{AC} - \widehat{DE})$          6. Substitution

*Proof continued on the following page*

**Part II**

Given: Secant $AB$ and tangent $BC$

Prove: $\angle B = \frac{1}{2}(\overset{\frown}{AC} - \overset{\frown}{CD})$

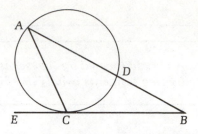

| | |
|---|---|
| **1.** Draw $AC$ | **1.** Construction |
| **2.** $\angle ACE = \angle A + \angle B$ | **2.** Exterior angle |
| **3.** $\angle B = \angle ACE - \angle A$ | **3.** Subtraction |
| **4.** $\angle ACE = \frac{1}{2}\overset{\frown}{AC}$ | **4.** $\angle$ formed by chord and tangent |
| **5.** $\angle A = \frac{1}{2}\overset{\frown}{DC}$ | **5.** Inscribed angle |
| **6.** $\angle B = \frac{1}{2}(\overset{\frown}{AC} - \overset{\frown}{DC})$ | **6.** Substitution |

**Part III**

Given: Tangents $AB$ and $BC$

Prove: $\angle B = \frac{1}{2}(\overset{\frown}{ADC} - \overset{\frown}{AC})$

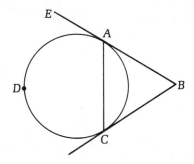

| | |
|---|---|
| **1.** Draw $AC$ | **1.** Construction |
| **2.** $\angle CAE = \angle ACB + \angle B$ | **2.** Exterior angle |
| **3.** $\angle B = \angle CAE - \angle ACB$ | **3.** Subtraction |
| **4.** $\angle CAE = \frac{1}{2}\overset{\frown}{ADC}$ | **4.** $\angle$ formed by chord and tangent |
| **5.** $\angle ACB = \frac{1}{2}\overset{\frown}{AC}$ | **5.** $\angle$ formed by chord and tangent |
| **6.** $\angle B = \frac{1}{2}(\overset{\frown}{ADC} - \overset{\frown}{AC})$ | **6.** Substitution |

Some examples:

**A secant and a tangent form an angle of 52°. The larger of the intercepted arcs is 190°. Find the smaller arc.**

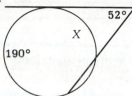

Since $52 = \frac{1}{2}(190 - X)$, $104 = 190 - X$, $X = 190 - 104$, $X = 86$. The smaller arc is 86°.

**Two tangents to a circle form an angle of 62°. Find the smaller intercepted arc.**

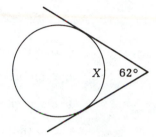

Here it was not necessary to have given the larger arc, since the smaller and larger arcs add to 360. The larger arc is $360 - X$. So $62 = \frac{1}{2}(360 - X - X)$, $124 = 360 - 2X$, $2X = 360 - 124$, $2X = 236$, $X = 118$. The smaller arc is 118°.

Notice that in this example, the smaller arc is supplementary to the angle formed by the two tangents. Is this always the case? Can we prove it?

**Given:** Tangents $PA$ and $PB$

**Prove:** $\angle P$ is supplementary to $\overarc{AB}$

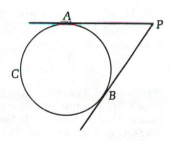

1. $PA$ and $PB$ are tangents

2. $\angle P = \frac{1}{2}(\overarc{ACB} - \overarc{AB})$

3. $\overarc{ACB} + \overarc{AB} = 360$

4. $\overarc{ACB} = 360 - \overarc{AB}$       You supply the reasons!

5. $\angle P = \frac{1}{2}[(360 - \overarc{AB}) - \overarc{AB}]$

6. $\angle P = \frac{1}{2}(360 - 2\overarc{AB})$
   $= 180 - \overarc{AB}$

7. $\angle P + \overarc{AB} = 180$

8. $\angle P$ is supplementary to $\overarc{AB}$

There is an interesting theorem about the lengths of a secant and a tangent to a circle that we will prove now, and then we will use it to solve a practical problem.

> *If a tangent and a secant are drawn to a circle from an external point, the tangent is the mean proportional to the secant and its external segment.*

Given: Secant $AC$, tangent $AB$

Prove: $\dfrac{AC}{AB} = \dfrac{AB}{AD}$

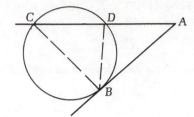

1. Draw $BC$, $BD$     1. Construction

2. $\measuredangle ABD = \frac{1}{2}\overset{\frown}{BD}$     2. $\measuredangle$ formed by tan. and chord

3. $\measuredangle ACB = \frac{1}{2}\overset{\frown}{BD}$     3. Inscribed angle

4. $\measuredangle ABD = \measuredangle ACB$     4. Substitution

5. $\measuredangle A = \measuredangle A$     5. Reflection

6. $\triangle ABD \sim \triangle ABC$     6. AA

7. $\dfrac{AC}{AB} = \dfrac{AB}{AD}$     7. Corr. parts of $\sim \triangle$ are prop.

Note that if $\dfrac{AC}{AB} = \dfrac{AB}{AD}$, $(AB)^2 = AC \times AD$, or $AB = \sqrt{AC \times AD}$.

This relationship is called the geometric mean. $X$ is the geometric mean to $Y$ and $Z$ when $X = \sqrt{YZ}$.

And now the problem:

**A picture is hung above eye level in a gallery. Where is the best place to stand to view it? (The same problem would apply to a screen in a drive-in movie, and many other situations.) If you stand too close or too far the viewing angle is decreased. The question becomes: Where does the eye have the maximum angle?**

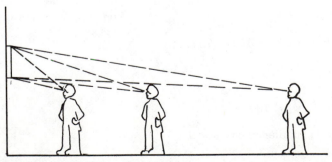

To eliminate the variable of the person's height let $AD$ = the height to the bottom of the picture, and let $AC$ = the distance of the top of the picture above eye level.

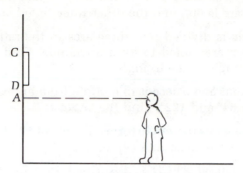

The desired point will now be where a circle drawn through the top and bottom of the picture is tangent to the line at eye level.

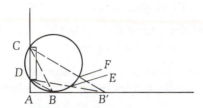

Since any other point, $B'$, would form two secants to the circle, it would have a smaller angle ($\angle B = \frac{1}{2}\overset{\frown}{CD}$ while $\angle B' = \frac{1}{2}(\overset{\frown}{CD} - \overset{\frown}{EF})$). Now, we use our theorem to determine where to stand.

The tangent, $AB$, is the mean proportion between the secant, $AC$, and its external segment, $AD$.

$$\frac{AC}{AB} = \frac{AB}{AD} \text{ or } AB = \sqrt{AC \times AD}$$

Therefore, stand at a distance which is the geometric mean between the height of the top of the picture and the height of the bottom of the picture above eye level.

It is interesting to note that this solution works even if the wall is not perpendicular.

Tile arcs in wall construction.

**EXERCISE SET 22**

1. Inscribed angles intercept arcs of 50°, 66°, 112°, and 125°. Find the number of degrees in each angle.

2. Two chords intersect within a circle to form an angle of 75°. If one intercepted arc is 60°, find the other intercepted arc.

3. A circle is divided into three arcs in the ratio of 2:3:5. The points of division are joined to form a triangle. Find the number of degrees in each angle of the triangle.

4. A tangent and a secant to a circle from an external point, P, intercept arcs of 84° and 122°. Find the angle at P.

5. Two tangents to a circle form an angle of 57°. Find the number of degrees in each intercepted arc.

6. Quadrilateral $ABCD$ is inscribed in a circle. $\angle A = 78°$, $\angle B = 85°$. Find $\angle C$ and $\angle D$.

7. Quadrilateral $PQRS$ is inscribed in a circle. $\overarc{PQ} = 72°$, $\overarc{PS} = 66°$, $\overarc{QR} = 108°$. Find each angle of $PQRS$.

8. The minor intercepted arc formed by two tangents to a circle is 70°. Find the angle between the tangents.

9. Two secants to a circle from an external point, P, form an angle of 36°. If the smaller intercepted arc is 53°, find the larger intercepted arc.

10. $\triangle DEF$ is an inscribed equilateral triangle. At $D$, a tangent is drawn to the circle. How large is the acute angle between the tangent and one side of $\triangle DEF$?

11. In the adjacent figure, $\overarc{AC} = 112°$, $\overarc{BD} = 84°$. Find the number of degrees in $\angle AED$.

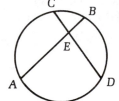

12. $\triangle ABC$ is isosceles, with $AC = BC$, $\angle A = 65°$. $BD$ is tangent to the circle at $B$. Find the number of degrees in $\angle ABD$.

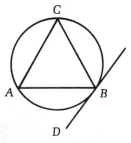

13. In the adjacent figure, $\overarc{CE} = 100°$, $\overarc{BD} = 30°$. Find the value of $\angle 1$, $\angle 2$, $\angle 3$, and $\angle 4$ ($\angle ABE$).

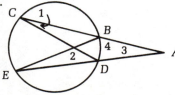

14. Prove that a trapezoid inscribed in a circle is isosceles.

15. △*ABC* is isosceles with *BA* = *BC*

Prove that *AD* = *CE*

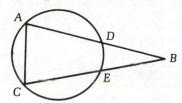

16. Construct a line segment equal to √12 cm.

17. Construct a line segment equal to √17 cm.

*18. *Prove:*  The opposite angles of an inscribed quadrilateral are supplementary.

*19. *Given:*  In circle *O*, chord *AB* is produced the length of a radius to point *C*. Line *CO* produced meets the circle at *D*.

*Prove:*  ∢*DOA* = 3∢*C*

*(Hint:*  ∢*DOA* = ∢*C* + ∢*A*, but ∢*A* = ∢*OBA*, and ∢*OBA* = ∢*C* + ∢*COB)*

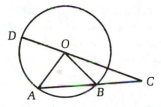

This problem is attributed to the Greek mathematician Archimedes (287–212 BC). This method allows us to trisect an angle!

Using the vertex of the angle we wish to trisect as center, draw a circle with any radius. Extend one side of the given angle through the circle. Mark off a segment equal to the radius on a straightedge. Now, with the straightedge on *A*, move it until one end of the marked segment is on the circle and the other end of the segment is on *DO* produced. These are points *B* and *C* respectively. ∢*ACD* will be ⅓ of ∢*AOD*. This technique does not contradict what we said earlier about the impossibility of trisecting an angle with straightedge and compass, since marking the distances on the straightedge and moving it until it "fits" by trial and error is not a legitimate construction method.

*20. Here is an interesting figure:

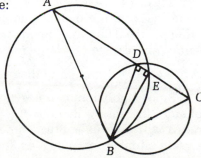

Since an angle inscribed in a semicircle is a right angle, ∢*AEB* is a right angle and ∢*BDC* is a right angle. Therefore, △*BDE* has two right angles! Can you explain this?

*21. One of the reasons the outside tires of a race car are changed more frequently in a race is the fact that the outside tires have to travel further. In the Indianapolis 500 the cars make 200 laps of the 2½ mile track. If the distance between inside and outside tires is 54 in., how much further do the outside tires travel during a race?

*22. Prove that in making a 90° turn, the outside tire of a car with width between tires of 6 ft. will travel more than 9 ft. further than the inside tire. Show that this fact does not depend upon the radius of the turn.

*23. Given two points, A and B, separated by an obstruction that does not permit a straightedge to be placed on it, construct portions of the line AB to the right of B and to the left of A.

A •      • B

## 23. The measure of a circle

Historically, a source of frustration to mathematicians was the attempt to relate the distance around a circle, the circumference, to the distance across that circle, the diameter. They realized that the size of the circle did not matter. They knew that the ratio of circumference to diameter is always the same. This constant ratio was given the name $\pi$, the Greek letter "pi," which is derived from the word "periphery," which means moving around, or the boundary of a rounded figure. We will adopt this *definition* also: **The ratio of the circumference of any circle to its diameter is the constant, pi.** $\frac{c}{d} = \pi$

For theoretical problems, the use of the symbol $\pi$ to represent this concept would be sufficient. Unfortunately, most of our tasks in actual situations are applied problems, and we need some numerical value for this ratio $\pi$. This is the problem that bothered the ancient mathematicians. They did not know that $\pi$ was an irrational number—that is, a non-repeating, non-terminating decimal. It cannot be represented as a fraction, an integer, or a terminating decimal. That means it is impossible to write its value in decimal notation! All we can do is use a symbol to express the concept.

So, what we need now is a decimal approximation to use in applied problems. There are several ways we can go about getting an approximation. We can use an inductive approach and very carefully measure some circles. If you have never done this, you should try it. With a piece of string or a flexible tape, measure the distance around a circular object. Then measure the distance across, and divide. Try this with several different-sized circles. Your answers should all be very close to the same value. An average of a large number of such experimental results will give a better *estimation* of the ratio $\frac{c}{d}$.

Since a polygon becomes more and more "circular" as the number of sides increases, the ratio of a polygon's perimeter to its diameter (2 radii) will give us another approach. Unfortunately there are only a certain number of polygons for which we know how to find the perimeter. With the use of trigonometry we can derive a formula for the perimeter of a polygon with a diameter of 1 (an easy number to divide by). The formula is $P = n$ sine $\frac{180}{n}$. In this formula, $n$ is the number of sides of the polygon and "sine" is a trigonometric function. Don't worry about where the formula came from or how to use it. This is only an example of a method for approximating pi. Using this formula with a pocket calculator, we get the following table of values.

| $n$ | $\frac{P}{d} \approx \pi$ | $n$ | $\frac{P}{d} \approx \pi$ |
|-----|------|------|---------|
| 3 | 2.59808 | 24 | 3.13263 |
| 4 | 2.82843 | 32 | 3.13655 |
| 5 | 2.93893 | 48 | 3.13935 |
| 6 | 3.00000 | 64 | 3.14033 |
| 7 | 3.03719 | 96 | 3.14103 |
| 8 | 3.06147 | 128 | 3.14128 |
| 9 | 3.07818 | 256 | 3.14151 |
| 10 | 3.09017 | 512 | 3.14157 |
| 12 | 3.10583 | 1024 | 3.14159 |
| 16 | 3.12145 | 2048 | 3.14159 |

These ratios are for an inscribed polygon.

Realize that each of these entries in the table has many more digits than the table shows, and has been rounded to 6 significant digits. All further entries, with greater values for $n$, will have 3.14159 for the first six digits. Therefore, the first six digits of pi must be 3.14159 (which is the case). A polygon drawn on a sheet of paper with 2048 sides would be so "round" it would be mistaken for a circle.

The entries used in the table are ones that might be constructed. Starting with a square and doubling the number of sides we would get 4, 8, 16, 32, . . . sides. Starting with a hexagon and doubling the number of sides we would get 6, 12, 24, 48, 96, . . . sides. Note that we could also approximate $\pi$ by using the *area* of an inscribed regular polygon. (Area $= \pi r^2$) However, it is interesting to see that to obtain the same degree of accuracy, we would have to use exactly twice as many sides.

The Rhind Papyrus indicates that the Egyptians, prior to 1700 BC, were calculating the volume of a spherical solid, using an approximation for pi of $3\frac{13}{81}$, which has a decimal equivalent of 3.16049 . . .

I *Kings* 7:23 and II *Chronicles* 4:2 are sometimes quoted to indicate that the Hebrews had lost any accuracy for the value of pi and were using 3 at the time of the building of Solomon's Temple in about 950 BC. Since these verses constitute a description, and not an architectural plan, it seems more reasonable to assume the dimensions given are approximations and not a reflection of the best known value. After all, the Hebrews had a long association with the Egyptians and should have learned their technology.

Archimedes, 212 BC, using the method of inscribing and circumscribing polygons, determined the value of pi to lie between $3\frac{10}{71}$ and $3\frac{1}{7}$; or using decimals, between 3.14085 . . . and 3.14286 . . . Check this with the value for 96 in our table. He used a polygon of 96 sides. By 150 AD Ptolemy was using a value of 3.14156 for pi. By 470 AD Tsu Chung-

**A BRIEF HISTORY OF THE ESTIMATION OF PI**

Chin (~429–500 AD) used $\frac{355}{113} \approx 3.14159292 \ldots$ but also showed that pi lies between 3.1415926 and 3.1415927!

In the seventeenth century much work was done in determining the value of pi. The technique at this time was the use of infinite series; that is, a series that goes on and on. The more terms one uses, the better the approximation becomes. Some of these series take a great many terms before the result is very accurate.

Wallis: $\quad \dfrac{\pi}{4} = \dfrac{2 \times 4 \times 4 \times 6 \times 6 \times 8 \times 8 \times 10 \times 10 \times \ldots \ldots}{3 \times 3 \times 5 \times 5 \times 7 \times 7 \times 9 \times 9 \times 11 \times 11 \times \ldots}$
(1655)

Lord Brouncker: $\quad \dfrac{\pi}{4} = \cfrac{1}{1 + \cfrac{1^2}{2 + \cfrac{3^2}{2 + \cfrac{5^2}{2 + \cfrac{7^2}{2 + \text{etc.} \ldots}}}}}$
(1658)

Leibniz: $\quad \dfrac{\pi}{4} = 1 - \dfrac{1}{3} + \dfrac{1}{5} - \dfrac{1}{7} + \dfrac{1}{9} - \dfrac{1}{11} + \dfrac{1}{13} - \text{etc.}$
(1673)

In the 1760s, J. H. Lambert proved that pi is irrational.

In 1874, pi was calculated by hand to 707 places.

In 1882, pi was proved to be transcendental as well as irrational by Lindemann, using a method devised by Hermite in 1873.

In 1956, pi was computed to 2035 places by one of the ENIAC computers. It took the machine 70 hours, a task which would require a lifetime of hard labor for someone computing by hand.

In 1966, pi was computed to one million digits (one wonders why). In the last few years, computers have become so fast and efficient that it is no great task to compute any desired number of digits of pi.

Just for curiosity's sake, pi to 110 decimal places is: 3.14159 26535 89793 23846 26433 83279 50288 41971 69399 37510 58209 74944 59230 78164 06286 20899 86280 34825 34211 70679 82148 08651

Now that we have the formula $C = \pi d$ or $C = 2\pi r$ we are able to do all sorts of computations and also are able to derive several other useful formulas. But before we derive more formulas, let us practice a bit with this one.

**What is the circumference of a circular track with a 32.00 m radius?**

$C = 2\pi r = 2\pi(32) = 64\pi$ m. If a decimal approximation is needed, you first have to decide how many digits are necessary, use one more digit than the desired number in the approximation of pi, and then round off to the correct number of digits. Assume we want the answer to the nearest centimeter. That means two decimal places. Use $\pi = 3.1416$ and $64\pi$ becomes 201.088 or 201 m 6 cm.

**What is the diameter of the largest circle that can be formed using 100 m of fencing?**

$$C = \pi d \text{ so } 100 = \pi d \text{ or } d = \frac{100}{\pi}$$

If a problem does not specify a desired accuracy, the answer should be left in exact form. That is, in terms of $\pi$ and radicals if they occur.

**Find the length of arc of a circle that is cut off by one side of a regular pentagon, if the radius of the circle is 7.**

$$C = 2\pi r = 2\pi(7) = 14\pi. \text{ The arc} = \frac{C}{5} = \frac{14\pi}{5}$$

**A 6 m by 8 m rectangular pool is to be fenced in with a circular fence. The fence must be at least 1 m away from the pool at all points. How much fencing is needed?**

First draw a sketch:

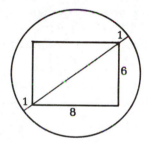

The diameter is the hypotenuse of a right triangle, with 2 additional meters. The triangle should be recognized as a triangle with 3-4-5 ratio, so the hypotenuse is 10, and the diameter is 12. $C = \pi d = 12\pi$ m. Give the answer to nearest decimeter. Let $\pi = 3.14$; $12\pi = 37.68$. So you need 37.7 m of fencing.

Here is a problem in which most people's intuition fails: assume that the Earth is a perfectly smooth sphere, and we stretch a string snugly around its circumference. We now cut the string and splice in an additional 1 meter of string so that it now fits loosely. If this slack amount is distributed evenly around the entire circumference, how far away from the surface would the string now stand? One centimeter? One millimeter? Enough to stick a piece of paper under? The surprising answer is about 16 cm, or 6 in.! If a string around a tin can, a coin, a tire, or any other circular object is increased by 1 meter, the amount of slack, if evenly distributed, will be this same 16 cm amount. The radius of the object involved is unimportant.

What we wish to find is the difference in the radii, R and r

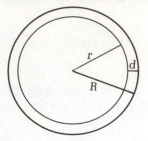

The circumference of the inner circle $C_r = 2\pi r$
The circumference of the outer circle $C_R = 2\pi r + 1$

The radius of the outer circle $R = \dfrac{C_R}{2\pi} = \dfrac{2\pi r + 1}{2\pi} = r + \dfrac{1}{2\pi}$

So the distance $d = R - r = \left(r + \dfrac{1}{2\pi}\right) - r = \dfrac{1}{2\pi} \approx .16$ m or 16 cm

If this still seems hard to believe, get a piece of string and try it with some different-sized objects.

Do you remember that postulate way back there at the beginning of the chapter that said: "Any property of regular polygons which does not depend on the number of sides of the polygon is also true of circles"? We called it the Limit Postulate and we are now ready to make use of it in finding a formula for the area of a circle.

---

> **The area of a circle is pi times the radius squared.** $A = \pi r^2$

Given: Circle O with radius r

Prove: $A = \pi r^2$

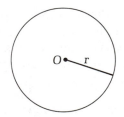

| | |
|---|---|
| **1.** Area of a polygon $= \frac{1}{2}ap$ | **1.** Area theorem |
| **2.** Area of the circle $= \frac{1}{2}ap$ | **2.** Limit Postulate |
| **3.** $a = r$ | **3.** Definition of apothem |
| **4.** $p = C$ | **4.** Definition of circumference |
| **5.** $A = \frac{1}{2}rC$ | **5.** Substitution |
| **6.** But $C = 2\pi r$ | **6.** Circumference formula |
| **7.** $A = \frac{1}{2}r(2\pi r) = \pi r^2$ | **7.** Substitution |

---

Since this is a very practical formula, let us practice its use.

**Find the area of a circular lawn 12 m in diameter.**

If the diameter = 12, r = 6 and $A = \pi r^2 = \pi 6^2 = 36\pi\,m^2$ ($m^2$ is read as square meters). For an answer to the nearest square meter let $\pi = 3.1$, $36\pi = 112\,m^2$

**What is the radius of the largest circle that can be planted with 16 kg of seed, if 1 kg of seed can cover 50 m²?**

16 kg will plant 50(16) = 800 m², so $A = 800$ in the formula $A = \pi r^2$; $A = 800$, so $800 = \pi r^2$, $r^2 = \dfrac{800}{\pi}$, $r = \sqrt{\dfrac{800}{\pi}} = \dfrac{20\sqrt{2}}{\sqrt{\pi}}$, or a radius of approximately 16 m.

 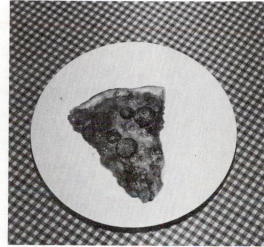

One of the more enjoyable examples of a sector.

There are some shapes associated with circles that it is sometimes useful to have a formula for. They are the sector, segment, and ring. A **sector** of a circle is that portion cut off by two radii and an arc.

A **segment** of a circle is that portion cut off by a chord and arc.

A **ring** is the area between two concentric circles.

We will now develop the formulas for the sector, segment, and ring.

*The area of a sector of a circle is the measure of its central angle divided by 360, times the area of the circle.*

$$A_{\text{sector}} = \frac{\text{central} \sphericalangle}{360} \pi r^2.$$

Given: Circle $O$ with sector $AOC$

Prove: $A_{\text{sector}} = \dfrac{\text{central} \sphericalangle}{360} \pi r^2$

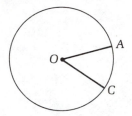

1. $\dfrac{A_{\text{sector}}}{A_{\text{circle}}} = \dfrac{\sphericalangle O}{360}$

2. $A_{\text{sector}} = \dfrac{\sphericalangle O}{360} A_{\text{circle}}$

3. $A_{\text{sector}} = \dfrac{\text{central} \sphericalangle}{360} \pi r^2$

1. The area is proportional to the central angle

2. Multiplication

3. Substitution

**Find the area of a sector of a circle with radius 12 and central angle of 60°.**

$$A = \frac{\text{cent.} \sphericalangle}{360} \cdot \pi r^2 = \frac{60}{360} \pi 12^2 = 24\pi$$

**What size circle is necessary to give a sector with a central angle of 50° an area of $\frac{4}{5}\pi$ square units?**

$$A = \frac{\text{cent.} \sphericalangle}{360} \cdot \pi r^2 \text{ so } \frac{4\pi}{5} = \frac{50}{360} \pi r^2, r^2 = \frac{144}{25}, r = \frac{12}{5}$$

> *The area of a segment of a circle is the area of its sector minus the area of its triangle.* (For the segment formed by the major arc and the chord, add the areas of the sector and the triangle.)

Given: Segment $AB$

Prove: $A_{\text{segment}} = A_{\text{sector } AOB} - A_{\triangle AOB}$

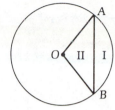

1. $A_{\text{sector } AOB} = A_{\text{I}} + A_{\text{II}}$       1. Whole is sum of its parts.

2. $A_{\text{I}} = A_{\text{sector } AOB} - A_{\text{II}}$       2. Subtraction

3. $A_{\text{segment}} = A_{\text{sector } AOB} - A_{\triangle AOB}$       3. Substitution

An example using this theorem:

**A chord cuts off a 120° arc of a circle with radius 7. Find the area of the segment formed.**

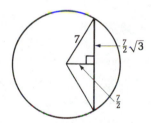

Draw the radii and a perpendicular to the chord. This will form a 30-60-90 triangle with sides $7$, $\frac{7}{2}$, and $\frac{7}{2}\sqrt{3}$  Using the formula:

$$A = \tfrac{120}{360}\pi\, 7^2 - \tfrac{1}{2}(7\sqrt{3})(\tfrac{7}{2}) = \tfrac{49}{3}\pi - \frac{49\sqrt{3}}{4}$$

> *The area of a ring is the difference of the areas of its circles.*
> $$A_{ring} = \pi R^2 - \pi r^2 = \pi(R^2 - r^2) = \pi(R - r)(R + r)$$

Given: Concentric circles with radii R and r

Prove: $A_{ring} = \pi(R^2 - r^2)$

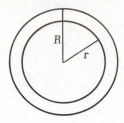

1. $A_{lg.\ circle} = \pi R^2$                    1. Area formula

2. $A_{sm.\ circle} = \pi r^2$                    2. Area formula

3. $A_{ring} = \pi R^2 - \pi r^2$                 3. Subtraction
      $= \pi(R^2 - r^2)$

**A circular lawn 20 m in diameter has a path 1 m wide around it. What is the area of the path?**

If $d = 20$, $r = 10$. The path is 1 so $R = 11$ and

$$A = \pi(R^2 - r^2) = \pi(R - r)(R + r)$$
$$A = \pi(11 - 10)(11 + 10) = 21\pi \text{ m}^2$$

Notice that it is often convenient to factor the difference of two squares—the arithmetic is easier.

**We have sufficient redwood chips to cover an area of $81\pi$ m². We want to lay out a circular path with them. What will be its inner radius? Make the path 1 m wide.**
Drawing a figure:

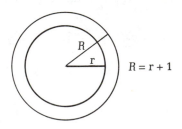

$R = r + 1$

$A = \pi(R^2 - r^2)$,   so $81\pi = \pi(r + 1)^2 - \pi r^2$,   $81 = r^2 + 2r + 1 - r^2$,
$81 = 2r + 1$,   $2r = 80$,   $r = 40$ m

**Problems involving linear measures.** Leave answers in exact form—that is, in terms of $\pi$ and radicals.

1. Find the circumference of a circle whose radius is 15 cm.

2. Find the radius of a circle whose circumference is 24 cm.

3. What is the radius of a circle if a central angle of 72° intercepts an arc of $20\pi$ cm?

4. Find the length of an arc intercepted by a central angle of 45° in a circle with radius 12 cm.

5. In a circle with radius of 6 cm an arc is $4\pi$ cm long. What is the degree measure of the arc?

6. What is the circumference of a circle if a central angle of 105° intercepts an arc of $14\pi$ cm?

7. Find the length of an arc of 30° in a circle with radius of 10 cm.

8. Find the number of degrees in the central angle of a circle with radius 8 cm, if the central angle intercepts an arc of 3 cm.

9. Find the circumference of the circle circumscribed about an equilateral triangle with a side of 6 cm.

10. Find the circumference of a circle circumscribed about a square with a side of $6\sqrt{2}$ cm.

11. Tangents to a circle of radius 4 cm form an angle of 120°. Find the perimeter of the figure bounded by the two tangents and the minor intercepted arc.

12. Two tangents are drawn to a circle of radius 5 cm and form an angle of 36°. Find the length of the major intercepted arc.

13. The minute hand of a clock is 12 cm long. How far does the end of the hand travel in 24 minutes?

14. What is the angle formed by the hands of a clock at 4:00?

*15. What is the angle formed by the hands of a clock at 5:05?

16. Two equal circles, each of radius 5 cm, intersect so that their common chord is equal to the radius. Find the perimeter of the figure bounded by the major arcs.

17. The length of an arc of a circle intercepted by an inscribed angle of 40° is 10 cm. Find the radius of the circle.

18. The apothem of a regular hexagon is $2\sqrt{3}$ cm. Find the circumference of the circumscribed circle.

19. The circumferences of two concentric circles are $20\pi$ and $14\pi$ cm. Find the width of the ring formed by the circles.

20. Find the perimeter of a sector of a circle of radius 10 cm if the angle of the sector is 66°.

21. A can of tennis balls holds three balls. Which is greater—the height of the can or its circumference? (You could probably win bets with this one!)

Which is greater—the height of the can or its circumference?

*22. The grooved music track of a record has an outer radius of 30 cm and an inner radius of 15 cm. The record revolves at a rate of $33\frac{1}{3}$ revolutions per minute and takes 20 minutes to play. Assume that the grooves are concentric circles and figure the length of the music track, if it were in a straight line.

*23. A rocket launched to the moon must "lead" its target in order to intercept it. If the rocket travels at 25,000 mph, the moon completes a circular orbit every 28 days, and the distance to the moon is 240,000 miles, what should be the "lead angle" $\theta$ at the time of launch?

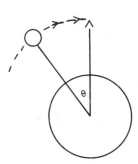

(Obviously not to scale!)

*24. Find the radius of the small circle.

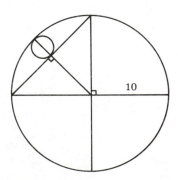

*25. Find the radius of the small circle.

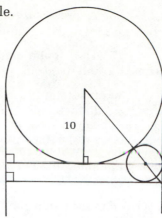

*26. Eratosthenes (280–195?BC) was not only the librarian of the world's greatest library, at Alexandria, but was also a great astronomer. He used the following method to determine the radius of the earth. He knew at the city of Syene there was a deep well whose waters were touched by sunlight at noon on the longest day of the year. So on that day, in Alexandria, he measured the angle of the sun from the vertical, and found the angle to be 7°12'. Knowing that the distance from Alexandria to Syene was 5000 stadia (a Greek unit of measure), he could then compute the radius of the earth.

What did he find as the radius (in stadia)?

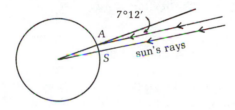

**Problems involving area measures.** Leave answers in exact form.

27. Find the area of a circle with a radius of 14 cm.

28. Find the radius of a circle with an area of $100\pi$.

29. Find the area of a circle with a circumference of $22\pi$.

30. Find the area of a sector, if its angle is 30° and radius is $2\sqrt{3}$.

31. Find the area of a sector, if its angle is 225° and its radius is $2\sqrt{14}$.

32. Find the angle of a sector, if the area of the sector is $28\pi$ and the radius of the circle is $2\sqrt{7}$.

33. The area of a sector is $4\pi$ sq. cm and its central angle is 30°. Find the radius of the sector.

34. Find the area of the smaller segment formed by a chord that is a side of an inscribed square in a circle of radius 3 cm.

35. Find the area of the smaller segment of a circle if the angular measure of the arc of the segment is 120°, and the radius of the circle is 10 cm.

36. A circle is inscribed in an equilateral triangle of side 12 cm. Find the area of the portion of the triangle which is outside the circle.

37. The circumference of a circle and the perimeter of a square are each equal to 20 cm. Which has the larger area and by how much?

38. Insulation is to be wrapped around a pipe 20 cm long. The pipe has a diameter of 21 cm, and 3 cm are needed to overlap for fastening. How many sq. cm of insulation are needed?

39. How much material is needed to construct a label for a can with a diameter of 8 cm and a height of 11 cm?

*40. A circle is inscribed in a square and an equilateral triangle is inscribed in the circle. Find the ratio of the area of the triangle to the area of the square.

*41. Here is an experimental method of finding an approximate value for pi: Find a floor with parallel lines equally spaced (like a hardwood floor). Make a batch of rods with the same length as the distance between the parallel lines on the floor (toothpicks or matches might do). Now throw the rods on the floor and count how many cross a line.

$$\pi \approx \frac{2 \text{ (total number of rods used)}}{\text{number of rods which cross a line}}$$

With an increased number of trials your approximation should become more accurate.

If you wish to use rods with a length shorter than the distance between the parallel lines, the formula gets more complicated:

$$\pi \approx \frac{2 \text{ (length of rod) (total number of rods used)}}{\text{(distance between lines) (number of rods which cross a line)}}$$

Try this experiment and report your results.

*42. Here is another method of experimentally approximating pi. Using a grid of squares and a set of disks, where the diameter of a disk is the same as the side of a square in the grid; throw the disks on the grid and count the number of them that cover a corner.

$$\pi \approx \frac{4 \text{ (number that cover a corner)}}{\text{total number of disks used}}$$

Checkers on a checker board or stones on a Go board should work well for this experiment. If the diameter of the disk is less than the side of the square, the ratio of $\frac{\text{number that cover a corner}}{\text{total number of disks used}}$ will approximate $\frac{\pi d^2}{4 \cdot l^2}$ where $d$ is the diameter of the disk, and $l$ is the side of the square.

Try this experiment and report your results.

*43. There is a formula for finding the length of a side of a regular polygon of $2n$ sides if the side of a polygon of $n$ sides is known.

$$S_{2n} = \sqrt{2r^2 - r\sqrt{4r^2 - (S_n)^2}}$$

where $r$ is the radius of the circumscribed circle, $S_n$ is the length of the side of the polygon with $n$ sides, and $S_{2n}$ is the length of the side of the polygon with $2n$ sides. A formula where one value $(S_n)$ can be used to find the next value $(S_{2n})$, and then used again to find $S_{4n}$, and so on, is called a **recursion** formula.

Start with a hexagon inscribed in a circle, with a radius of $1 (S_6 = 1)$ and compute $S_{96}$ with the formula, finding approximate values for $\pi$ as you go. The following chart might help organize the data:

| $n$ | $S_n$ | Perimeter $n(S_n)$ | $\pi$ Approximation Perimeter/$2r$ |
|---|---|---|---|
| 6 | 1 | 6 | 3.00000 |
| 12 | | | |
| 24 | | | |
| 48 | | | |
| 96 | | | |

It is believed that Archimedes approximated $\pi$ by using an inscribed polygon of 96 sides. (He didn't have a pocket calculator either—no pockets!)

*44. Start with a square inscribed in a circle with radius of 1, and approximate $\pi$ with a polygon of 4096 sides. (Use the recursion formula from problem 43.) Watch out for round-off and truncation errors.

**45. Write a computer program with a loop which will use the recursion formula in problem 43 to approximate $\pi$ to any required number of significant digits.

*46. All that is needed to derive the recursion formula of problem 43 is the Pythagorean Theorem and some Algebra (although perhaps more than you learned in Elementary Algebra). Derive the formula.

This figure might help to get you started:

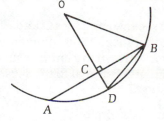

$OB = R$
$AB = S_n$
$BD = S_{2n}$

## 6–Summary

Theorems about circles:

Any three noncollinear points determine a circle.

In the same circle or in equal circles, equal arcs have equal central angles.

In the same circle or in equal circles, equal central angles have equal arcs.

In the same circle or in equal circles, equal arcs have equal chords.

In the same circle or in equal circles, equal chords have equal arcs.

In the same circle or in equal circles, equal central angles have equal chords.

In the same circle or in equal circles, equal chords have equal central angles.

A line through the center of a circle, perpendicular to a chord, bisects the chord and its arcs.

The perpendicular bisector of a chord passes through the center of the circle.

In the same circle or in equal circles, equal chords are equidistant from the center.

In the same circle or in equal circles, chords equidistant from the center are equal.

The perpendicular to a radius at its extremity is a tangent.

A radius drawn to the point of contact of a tangent is perpendicular to the tangent.

Tangents to a circle from an external point are equal.

Two parallel lines intercept equal arcs on a circle.

Theorems about angles and arcs:

An inscribed angle is half the measure of its intercepted arc.

Angles inscribed in the same arc or in equal arcs are equal.

An angle inscribed in a semicircle is a right angle.

An angle formed by two intersecting chords is half the sum of the two intercepted arcs.

An angle formed by a chord and a tangent at one end of the chord is half its intercepted arc.

An angle formed by two secants, by a secant and a tangent, or by two tangents is half the difference of the intercepted arcs.

If a tangent and a secant are drawn to a circle from an external point, the tangent is the mean proportional to the secant and its external segment.

Theorems about the measure of a circle:

The ratio of the circumference of any circle to its diameter is the constant, pi. (Definition)

The area of a circle is pi times the radius squared.

$$A = \pi r^2$$

The area of a sector of a circle is the measure of its central angle divided by 360, times the area of the circle.

$$A = \frac{C}{360}\,\pi r^2$$

The area of a segment of a circle is the area of its sector minus the area of its triangle. (For the segment formed by the *major* arc and the chord, *add* the areas of the sector and the triangle.)

The area of a ring is the difference of the areas of its circles.

*"Study of Angles"* by George Alves.

# CHAPTER 7

# SURFACE AND VOLUME

In this chapter, we extend our geometry from the plane into three-dimensional space. We want to develop some theorems and formulas about the surface area of solids, and the volume of solids. Of course there are many, many different three-dimensional shapes, so we will have to carefully define the special few for which we can develop formulas.

A final section, with a great deal of application in the natural world about us, can be included to conclude this chapter. The golden mean is a ratio which seems to pop up in the most unexpected places.

---

The concept of the area of a plane surface is easily extended to the surface of a solid that may or may not be "flat". If a sheet of paper is taped end-to-end, it forms the lateral (side) surface of a cylinder. If the surface of a sector has its radii taped together we get the lateral surface of a cone.

So, to find the surface area of a given solid, we try to "unfold" the solid in such a manner that we have shapes for which we have formulas.

## 24. Surface of solids

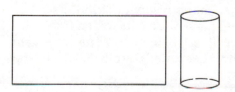
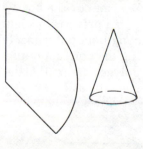

Sometimes this is fairly easy, as in the cylinder, cone, or pyramid; and sometimes it is very difficult, as in the sphere. What we will do now is try to analyze the surface of some solids and derive some formulas.

A rectangular solid is any kind of a "box," in which each of the six faces is a rectangle, with the opposite sides of the solid being congruent and parallel. The proper name for such a solid is a rectangular parallelepiped.

The surface, S, of a rectangular parallelepiped is therefore:

$$S = 2 \text{ sides} + 2 \text{ ends} + \text{top} + \text{bottom}$$

If the dimensions of length, width, and height are given,

$$S = 2(lh) + 2(wh) + 2(lw)$$

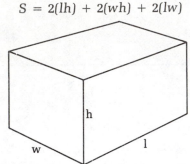

Since a cube is a special case of the above formula, and in a cube $l = w = h$, the **surface of a cube**, $S = 6a^2$, where $a$ is the measure of any edge.

How much cardboard is required to make a box measuring 40 cm by 60 cm by 1 m? (neglecting necessary overlap at edges)

$$S = 2(lh) + 2(wh) + 2(lw)$$

It is unimportant which dimensions we call $l$ or $w$ or $h$, but it is necessary to use like measurements. Let's do it in meters.

$$S = 2(.4)(.6) + 2(1)(.6) + 2(.4)(1)$$

$$S = .48 + 1.2 + .8 = 2.48 \text{ m}^2$$

How large a cubical box can be covered with 2400 cm² of wrapping paper?

$$S = 6a^2$$
$$6a^2 = 2400$$
$$a^2 = 400$$
$$a = 20$$

We can cover a box with edge of 20 cm.

A **prism** is the figure formed when the corresponding vertices of two congruent polygons, lying in parallel planes, are joined. The lines joining the corresponding vertices are called lateral edges. The congruent polygons are called the bases, and the other surfaces are called the lateral faces, or as a group, the lateral surface.

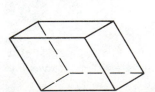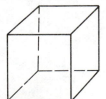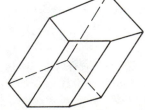

The prisms are named by the figure in the bases. The first two on the opposite page are quadrilateral prisms, while the third is a pentagonal prism. If the lateral edges are perpendicular to the plane of the bases, the prism is a **right prism;** otherwise, it is an **oblique prism.**

This Pacific Gas & Electric tank for storage of natural gas has so many sides it appears cylindrical. It is a prism with 28 sides. It is 379 ft. high and 254 ft. 4 in. in diameter. It holds 17,066,900 cu. ft. of gas (liners on the interior take up some of the volume).

Here is a right regular hexagonal prism:

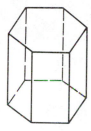

And an oblique triangular prism:

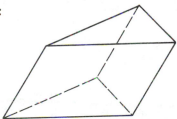

For the surface of an oblique prism, we simply add the surfaces of all the lateral faces, which will be parallelograms, and the two congruent bases, which will be polygons (not necessarily regular). For the surface of a right prism, we can do better. Since each lateral face will be a rectangle with the height of the prism as one of its dimensions, and the side of the base polygon as the other dimension, the sum of the lateral faces will be the perimeter of the base times the height. So the lateral surface, $L = ph$. If we wish the total surface, simply add the two bases. **The total surface of a right prism, $S = ph + 2B$,** where $p$ is the perimeter of the base, $h$ is the height of the prism, and $B$ is the area of one of the bases.

**Find the lateral surface of a regular right pentagonal prism with edges of 6 cm and height of 20 cm.**

$$L = ph = (5)(6)(20) = 600 \text{ cm}^2$$

**Twenty-four hexagonal columns 1 m in diameter (vertex to vertex) and 6 m tall are to be painted on all surfaces with a paint which covers 10 m² per liter. How much paint is needed?**

$$S_{\text{of 1 column}} = ph + 2B; \text{ so we need to find } p \text{ and } B.$$

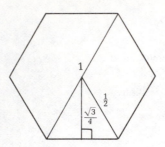

$d = 1$, so $r = \frac{1}{2}$, which means each side $= \frac{1}{2}$ and $p = 6(\frac{1}{2}) = 3$. For $B$, the area of a polygon is $\frac{1}{2}$(apothem)(perimeter), which in this case is

$$\frac{1}{2}\left(\frac{\sqrt{3}}{4}\right)(3) = \frac{3\sqrt{3}}{8}$$

Therefore

$$S = (3)(6) + 2\left(\frac{3\sqrt{3}}{8}\right) = 18 + \frac{3\sqrt{3}}{4}$$

For 24 columns, $S = 432 + 18\sqrt{3}$ m².

Dividing by 10, the coverage of each liter, we get $43.2 + 1.8\sqrt{3}$ liters; or for a decimal approximation to the nearest tenth of a liter, 46.3 liters of paint are needed.

Now, since we have a postulate that says "Any property of regular polygons which does not depend on the number of sides of the polygon is also true of circles," we can extend our formula for the surface of a prism to the surface of a cylinder. A **cylinder** is the solid formed when the corresponding parts of two congruent circles, lying in parallel planes, are joined—forming a curved lateral surface.

If the surface of a right prism is $S = ph + 2B$, then considering the circular base of a right cylinder as just a polygon with an infinite number of sides, the surface of a right cylinder becomes: $S = (2\pi r)h + 2(\pi r^2)$ or if you wish to factor,

$$S = 2\pi r(h + r)$$

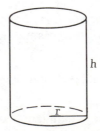

Consider the oblique circular cylinder. Since it is a circular cylinder, the two bases are circles the same as before. But what happens to the lateral surface? What do you get if you "unroll" an oblique circular cylinder?

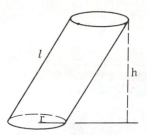

The shape is not easy to visualize or construct, but it will look like this:

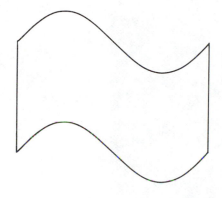

A sculpture on the surface of a cube.
*Stanford Museum.*

The interesting thing about it, although we can not prove it here, is that it has exactly the same area as the lateral surface of the right circular cylinder with the same base. It is much the same relationship as that of the area of a rectangle and a parallelogram that have the same base and height. As the parallelogram (or cylinder) is "pushed over", its one dimension (slant height) becomes greater, but the other dimension (circumference of a right cross section) becomes less.

The Leaning Tower of Pisa is not an oblique circular cylinder, but is a right circular cylinder which has been tipped over. *Photo Scala, Florence.*

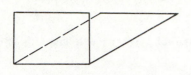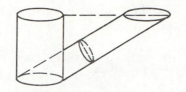

Since the area of the rectangular surface of the right cylinder is the same as the area of the surface of the oblique cylinder, the formula for the total surface will be the same—$S = 2\pi r(h + r)$. When using this formula for an oblique cylinder, be sure to use $h$, the perpendicular distance between the two bases, and not $l$, the slant height.

**What is the area of the label on a can that has a height of 14 cm and a diameter of 8 cm?**

The lateral surface is $L = 2\pi rh$, so

$$L = 2\pi\,4(14) = 112\pi\,\text{cm}^2$$

**How much tin is required to construct the can described above?**

Total surface $= 2\pi r(h + r)$

$$S = 2\pi\,4(14 + 4) = 144\pi\,\text{cm}^2$$

A **pyramid** is the solid formed when each vertex of a polygon is joined to some point not in the plane of the polygon.

Pyramids are named in the same manner as prisms, by the shape of the base and as either right or oblique. In order for a pyramid to be a right pyramid, its vertex must be directly above the center of the base. Illustrated here, from left to right, are a right hexagonal pyramid, a quadrangular pyramid, and an oblique pentagonal pyramid.

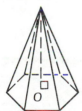

The surface of an oblique pyramid is composed of a set of triangular faces and a polygonal base. Each of the triangular faces might have a different height (called the slant height of the pyramid). Therefore, a great deal of information is needed to compute S. For the regular right pyramid, all of the faces will be congruent and each face is a triangle with area of $\frac{1}{2}bh$. The sum of all the bases of the triangular faces is the perimeter of the base polygon, and the height of each face is the slant height, $l$.

So, the surface of a regular right pyramid is

$$S = \tfrac{1}{2}pl + B$$

where $p$ is the perimeter of the base, $l$ is the slant height, and $B$ is the area of the base polygon.

The slant height, $l$, can often be found by using other given information. For instance:

**Find the surface of a regular right hexagonal pyramid with base side of 6 cm and height of 10 cm.**

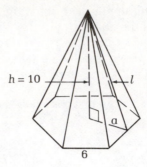

In order to find $l$ we use the right triangle formed by $l$, $h$, and $a$, the apothem of the hexagon.

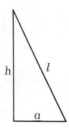

In the base, $a$ is the altitude of an equilateral triangle with side of 6, so $a = 3\sqrt{3}$

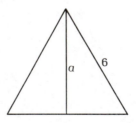

Now, go back to the right triangle, $l^2 = h^2 + a^2 = 10^2 + (3\sqrt{3})^2 = 127$ so $l = \sqrt{127}$. Now apply the formula for total surface.

$$S = \tfrac{1}{2}pl + B = \tfrac{1}{2}(36)(\sqrt{127}) + \frac{6^3\sqrt{3}}{4} = 18\sqrt{127} + 54\sqrt{3} \text{ cm}^2$$

A **cone** is the solid formed when each point of a circle is joined to a point not in the plane of the circle—forming a curved lateral surface.

Now let us consider the surface of a cone. Finding the lateral surface of an oblique cone is more than we can attempt here, but the right circular cone can be considered as a right regular pyramid with an infinite number of sides. Remember that theorems about polygons that do not specify the number of sides also apply to circles.

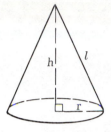

right circular cone          oblique circular cone

So, $S = \frac{1}{2}pl + B$ where $p$ will be the circumference of a circle, and $B$ is the area of a circle, and $l$ is the slant height.

$$S = \frac{1}{2}(2\pi r)l + \pi r^2$$
$$S = \pi rl + \pi r^2$$
$$S = \pi r(l + r)$$

Again, notice that if $r$ and $h$ are given, that $l$ can be found by the use of the Pythagorean Theorem.

**Find the lateral surface, $L$, and the total surface, $S$, of a right circular cone with radius of 6 cm and height of 8 cm.**

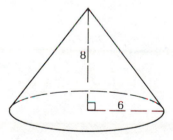

$S = \pi r(l + r)$ so we need the value of $l$. In the right triangle with legs of 6 and 8, the hypotenuse, $l$, will be 10. (This is the 3-4-5 right triangle if you recognize it, or if not, use the Pythagorean Theorem.)

$$S = \pi(6)(10 + 6) = 96\pi \, cm^2$$
$$L = \pi rl = \pi(6)(10) = 60\pi \, cm^2$$

A **sphere** is the set of points a given distance (radius) from a given point (center). The area of the surface of a sphere is a little more difficult to derive, so don't be discouraged if you get lost along the way.

A **zone** is the surface of a sphere cut off by two parallel planes. We will divide the total surface into a number of strips, called zones (like the Arctic Zone, the Temperate Zone, and so forth), find the area of each zone, and then add the areas of all the zones. But we don't have a formula for the area of a zone, so we approximate it as the area of a portion of the surface of a cone.

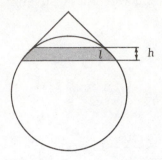

In the figure, $l$ is the slant height of the surface of the cone, and $h$, the perpendicular distance, is the height of the zone. This portion of a cone is called a frustum of a cone.

A **frustum of a cone** is the portion of a cone cut off by two parallel planes. The plan is to derive a formula for the frustum of a cone, and since as $h$ gets smaller and smaller, the area of the frustum of a cone approaches the area of the zone, we can then find the sum of the areas of all the zones to get the area of the sphere.

The surface of the frustum of a cone is equal to the lateral surface of the large cone, minus the lateral surface of the small cone.

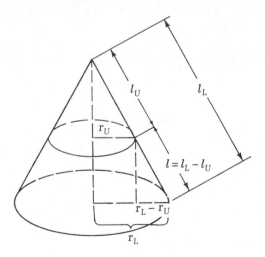

$$S = S_L - S_U = \pi r_L l_L - \pi r_U l_U$$

substituting $l_U = l_L - l$ we get

$$S = \pi r_L l_L - \pi r_U (l_L - l)$$

doing some factoring gives

$$S = \pi [(r_L - r_U)l_L + r_U l]$$

now by similar triangles in the above figure, $\dfrac{r_L}{l_L} = \dfrac{r_L - r_U}{l}$ which gives

us $l_L = \dfrac{r_L l}{r_L - r_U}$, which we substitute, giving

$$S = \pi(r_L l + r_U l)$$

$$S = \pi l(r_L + r_U)$$

Now we will make use of a "mid-radius" which will be the average of the upper and lower radii of the frustum of the cone.

$$r_M = \frac{r_L + r_U}{2} \text{ or } r_L + r_U = 2r_M$$

so

$$S = 2\pi r_M l$$

In looking again at our zone, we find some similar triangles.

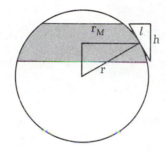

$\dfrac{r_M}{r} = \dfrac{h}{l}$ or $r_M l = rh$, which we substitute into $S = 2\pi r_M l$ giving

$$S = 2\pi rh$$

If we divide the surface of the sphere into a lot of very small zones, then the surface of all the frustums of cones will approach the surface of all the zones, which is the surface of the sphere. The sum of the heights of all the zones will be the height of the sphere, which is $2r$. So our formula becomes

$$S = 2\pi r(2r)$$

$S = 4\pi r^2$—The formula for the surface of a sphere.

**What is the surface area of a spherical butane tank which has a diameter of 1.5 m?**

$S = 4\pi r^2$, so we need $r$; if $d = 1.5$ then $r = .75$

$$S = 4\pi(.75)^2 = 2.25\pi \, \text{m}^2.$$

The rate at which an object melts depends, among other things, upon the exposed surface area. What is the surface area of a hemisphere of ice cream on a cone with a diameter of 6 cm?

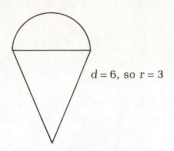

$d = 6$, so $r = 3$

$$S = \tfrac{1}{2}(4\pi r^2) = \tfrac{1}{2}(4\pi\, 3^2) = 18\pi \, \text{cm}^2$$

(Now we have some pi to go with our ice cream!)

### A Summary of Surface Formulas

| Surface of a | equals |
|---|---|
| rectangular parallelepiped | $2(lh) + 2(wh) + 2(lw)$ |
| cube | $6a^3$ |
| prism (right or oblique) | $ph + 2B$ |
| circular cylinder (right or oblique) | $2\pi r(h + r)$ |
| pyramid (right regular) | $\tfrac{1}{2}pl + B$ |
| right circular cone | $\pi r(l + r)$ |
| frustum of a cone | $2\pi r_M l$ |
| sphere | $4\pi r^2$ |

**EXERCISE SET 24**

1. What is the total surface area of a room 8 m by 10 m by 3 m?

2. Find the surface of a regular tetrahedron with an edge of 12 cm. (A regular tetrahedron is a pyramid with equilateral triangles for *all* of its faces.)

3. A semicircle of radius 20 cm is folded to form a cone (dunce cap?). What will be its height?

4. A soda can has a diameter of 6.5 cm and a height of 12 cm. How much aluminum is required to make the can?

5. A right hexagonal prism with a side of 2 cm is 15 cm tall. Find its lateral surface.

6. Find the total surface for the prism in problem 5.

7. A spherical propane tank is 2 m in diameter. What is its surface?

8. The edge of the base of a regular right hexagonal prism is 4 cm, and the prism has a height of 9 cm. Find the lateral area.

9. The lateral area of a regular right square prism is 96 sq. cm, and the altitude is 8 cm. Find the length of a base edge.

10. The base of a right parallelepiped is a rhombus with diagonals of 10 cm and 24 cm. The altitude of the parallelepiped is 7 cm. Find the total area. (A parallelepiped is a prism which has parallelograms for all of its faces.)

11. The lateral area of a right circular cylinder is 132 cm² and the circumference of the base is 12 cm. Find the altitude.

12. The total area of a right circular cylinder is $56\pi$ cm ² and the altitude is 12 cm. Find the radius of the base.

13. The radius of the base of a right circular cylinder is 8 cm and the altitude is 15 cm. A right circular cone is placed inside the cylinder so that their bases coincide and the vertex of the cone just touches the upper base of the cylinder. What is the ratio of the lateral areas of the cone and the cylinder?

14. A cylinder with a base radius of 5 cm and an altitude of 24 cm is inscribed in a sphere. Find the surface area of the sphere.

15. The total surface area of a cube is 294 cm². Find the length of an edge.

16. A pyramid with an equilateral triangle for its base has lateral edges of 8 cm, and each edge of the base measures 6 cm. Find the lateral area. Find the total area.

17. A regular right hexagonal prism has a base area of $6\sqrt{3}$. If the lateral faces are rectangles twice as tall as they are wide, find the lateral area.

18. Find the total surface area of a regular icosahedron which has edges of 5 cm. (see page 321)

19. A 3-4-5 right triangle is rotated about the 3 unit leg. What is the lateral surface of the cone generated?

20. A 3-4-5 right triangle is rotated about its hypotenuse. What is the total surface area of the solid generated?

21. What is the total surface area of a box 20 × 30 × 50 cm?

22. What is the area of the label of a can 15 cm high and 8 cm in diameter?

23. What is the total surface of the can above?

24. Using equilateral triangles 3 cm on an edge, what are the ratios of the areas of the regular tetrahedron and octahedron? (see page 321)

25. A regular right hexagonal pyramid has a side of the base 8 m and a height of 12 m. Find the total surface area.

26. Find the total surface of a right circular cone with base radius of 4 cm and slant height of 8 cm.

27. Find the total surface of a right circular cone with base radius of 5 cm and height of 8 cm.

28. Find the lateral surface of the frustum of a cone that has lower base radius of 8 cm, upper base radius of 4 cm, and slant height of 10 cm.

29. A large sphere on the top of a flagpole is to be goldleafed. Find its total surface if its diameter is 25 cm.

30. The cost of goldleafing the above sphere was too great, so it was decided to goldleaf only the bottom 15 cm (no one but the birds will see the top). What will be the surface of this zone?

## 25. Volume of solids

We now turn our attention to the volume of solids. Just as area is thought of as the number of square units that can be contained within a two-dimensional boundary, the volume of a solid can be thought of as the number of cubic units that will fit within the boundary of a three-dimensional solid. And, since we began our development of the area formulas by defining the area of a rectangle as $A = lw$, we begin our development of the volume formulas by defining the volume of a rectangular solid as

$$V = lwh$$

Consider the rectangular solid with dimensions of 3, 4, and 5 units (what unit you use is unimportant).

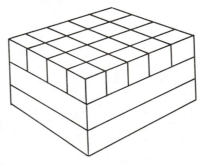

The area of the base is 20 square units, so 20 unit cubes can be placed on the bottom. Now since the height is 3 units we can make 3 layers of cubic units, 20 in each layer, and the volume is $V = 3(20)$ or $3(5)(4)$ or $V = lwh$. Of course, this is only an intuitive approach to our definition, and not a proof.

As in our definition of area, we have problems with dimensions that don't "fit" nicely and with irregular boundaries. We solve these problems in the same manner—by using smaller and smaller units and eventually "filling" the solid with an infinite number of infinitely small unit cubes. This technique is formalized in calculus and is called a Riemann Sum (remember him?). We will accept, on the basis of our intuition, that $V = lwh$ for all sorts of dimensions, rational or irrational.

Now since a cube is a rectangular solid in which $l = w = h$, we can derive the **volume of a cube** as $V = a^3$ where $a$ is the measure of any edge.

Suppose now that we take our rectangular solid and "push it out of shape," so that the faces are parallelograms rather than rectangles. This kind of solid is called a parallelepiped. The opposite faces are parallelograms and are in parallel planes.

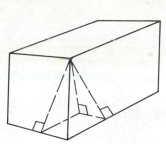
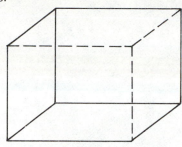

If the base is not a rectangle we can still find the volume in the same manner as before. We place as many cubic units on the base as there are square units of area in the base, and then make as many layers as there are units of height. The volume is the area of the base, $B$, times the height, $h$,

$$V = Bh$$

This concept can also be used for the prism and the cylinder, whether regular or not, right or oblique.

In the special case of the circular cylinder the base is the area of a circle, $\pi r^2$, so $V = Bh$ becomes

$$V = \pi r^2 h$$

**Find the volume of an oblique hexagonal prism with edge of the base 7 cm and height of 12 cm.**

$B = \frac{1}{2}ap$, where $a$ is the apothem of a hexagon with side 7, and $p = 6(\text{sides}) = 6(7) = 42$.

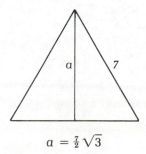

$a = \frac{7}{2}\sqrt{3}$

$V = Bh = \frac{1}{2}aph = \frac{1}{2}(\frac{7}{2}\sqrt{3})(42)(12) = 882\sqrt{3}$ cc (cc is cubic cm)

The Transamerica Building in San Francisco is a modern-day example of a pyramid. *Courtesy Transamerica Corporation.*

**If the circumference of a tennis ball is 21 cm, find to the nearest tenth of a cc the volume of a can holding three tennis balls.**

Here, circumference is given and we need radius, so, $C = 2\pi r$ gives $r = \dfrac{C}{2\pi}$. We also need $h$ for our formula, and the height of the can will be three diameters or 6 radii or $\dfrac{3C}{\pi}$

$$V = \pi r^2 h = \pi\left(\frac{21}{2\pi}\right)^2\left(\frac{3C}{\pi}\right) = \pi\left(\frac{21}{2\pi}\right)^2\left(\frac{3 \times 21}{\pi}\right) = \frac{27\ 783}{4\pi^2}$$

Using three digits for $\pi$; $V = \dfrac{27\ 783}{4(3.1416)(3.1416)} = 703.7$ cc to nearest tenth.

The volume of a pyramid is not easily derived. We will indicate the method, but will not attempt the proof. By passing two planes through certain edges of a triangular prism, it can be shown that the planes divide the prism into three equal pyramids. These pyramids are equal in volume, but are not congruent to each other. Thus, the volume of a triangular prism is one-third of the volume of the prism having the same base and height. So our formula for the volume of a pyramid is

$$V = \tfrac{1}{3}Bh$$

Since any pyramid can be divided into triangular pyramids, this formula applies to the volume of any pyramid, right or oblique, with any kind of base!

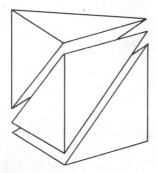

**Find the volume of the stone used in the construction of the Cheops Pyramid if the original dimensions were a square base of 756 feet on a side and a height of 481 feet.**

$$V = \tfrac{1}{3}Bh = \tfrac{1}{3}(756)^2(481) = 91\ 600\ 000 \text{ cubic feet.}$$

(Notice that our answer is rounded to three significant digits, because our original data only gives us three digit accuracy. No answer can be any more accurate than the original measurements.)

Once again, we will consider the cone as a pyramid with an infinite number of sides, and apply the Limit Postulate. Therefore, the volume of any cone, right or oblique, is $V = \tfrac{1}{3}Bh$. If we have a circular cone, then the base is the area of a circle, $B = \pi r^2$, and the volume of the circular cone is

$$V = \tfrac{1}{3}\pi r^2 h$$

What volume of ice cream can be packed into an ice cream cone which measures 6 cm across the top and is 12 cm high?

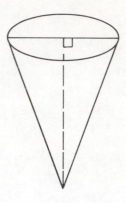

The diameter is 6 cm, so the radius is 3 cm.

$$V = \tfrac{1}{3}\pi r^2 h = \tfrac{1}{3}(\pi)(3)^2(12) = 36\pi \text{ cc}$$

To find the volume of a sphere, we proceed as follows: consider the sphere to be composed of a great many very small pyramids, all with their vertices at the center of the sphere. As the number of pyramids increases, the sum of their bases approaches being the surface of the sphere, and the height of the pyramids approaches the radius of the sphere.

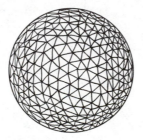 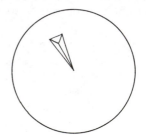

$$V = \tfrac{1}{3}B_1 h + \tfrac{1}{3}B_2 h + \tfrac{1}{3}B_3 h + \ldots$$

$$V = \tfrac{1}{3}h(B_1 + B_2 + B_3 + \ldots)$$

but in the limiting case, $B_1 + B_2 + B_3 + \ldots = S$ and $h = r$, so

$$V = \tfrac{1}{3}rS = \tfrac{1}{3}r(4\pi r^2)$$

$V = \tfrac{4}{3}\pi r^3$—The formula for the volume of a sphere.

**Find, to the nearest cc, the volume of a basketball, if the basketball has a circumference of 76 cm.**

The formula for the volume, $V = \tfrac{4}{3}\pi r^3$, uses r, so we need the radius instead of the circumference. But $C = 2\pi r$ so $r = \dfrac{C}{2\pi}$

$$V = \frac{4}{3}\pi\left(\frac{C}{2\pi}\right)^3 = \frac{4}{3}\pi\left(\frac{76}{2\pi}\right)^3 = 7400 \text{ cc (2 digit accuracy).}$$

The volume of the sphere is twice the volume of the cone, and the volume of the cylinder is three times the volume of the cone. The volume of the sphere and the cone combined is exactly equal to that of the cylinder.

A very interesting relationship develops if we look at the volumes of a right circular cylinder, a sphere, and a right circular cone. $V_{cylinder} = \pi r^2 h$, but if its height is equal to its diameter, $2r$, $V_{cylinder} = 2\pi r^3$. But notice that the volume of the sphere is exactly $\frac{2}{3}$ of the volume of the cylinder. $\frac{2}{3}(2\pi r^3) = \frac{4}{3}\pi r^3$ ($V_{sphere} = \frac{4}{3}\pi r^3$); and furthermore, the volume of the cone with the same height is exactly half that of the sphere (and one-third that of the cylinder). $V_{cone} = \frac{1}{3}Bh$ where $B = \pi r^2$ and $h = 2r$, so $V_{cone} = \frac{1}{3}(\pi r^2)(2r) = \frac{2}{3}\pi r^3$. The difference in volume between a sphere and its circumscribed cylinder is exactly equal to the volume of the inscribed cone in that same cylinder. A cylinder of water would exactly fill the sphere and the cone!

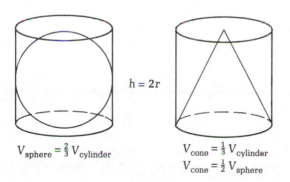

$$h = 2r$$

$$V_{sphere} = \tfrac{2}{3} V_{cylinder}$$

$$V_{cone} = \tfrac{1}{3} V_{cylinder}$$
$$V_{cone} = \tfrac{1}{2} V_{sphere}$$

Another item that should be mentioned when discussing solids, and which is a linear measure rather than a surface or a volume, is the space diagonal of a rectangular solid.

319

In the illustrated rectangular solid:

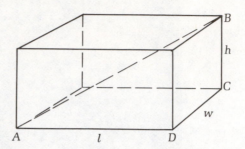

*AB* is the space diagonal. It is the greatest interior dimension of a rectangular solid. Its length is the length of the longest object that will fit into a "box". To find its length, *d*, in terms of the dimensions of the solid, we need to use two right triangles. First find the length of the diagonal of the base. In triangle *ACD*

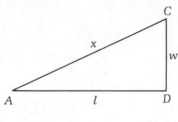

$$x^2 = l^2 + w^2 \text{ or } x = \sqrt{l^2 + w^2}$$

Next use the right triangle *ABC* to find the length of *d*.

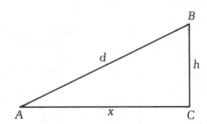

$$d^2 = x^2 + h^2, \ d^2 = (\sqrt{l^2 + w^2})^2 + h^2, \ d^2 = l^2 + w^2 + h^2$$

therefore,

$$d = \sqrt{l^2 + w^2 + h^2}$$

**Find the length of the longest pair of skis that will fit in a closet measuring 50 cm by 80 cm by 2.00 m.**

$$d = \sqrt{50^2 + 80^2 + 200^2} = 10\sqrt{5^2 + 8^2 + 20^2} = 10\sqrt{489}$$

or to the nearest cm, 221 cm.

One final topic of interest should be mentioned in our discussion of the measure of three-dimensional objects. In plane geometry there is an infinity of regular polygons: the equilateral triangle, the square, the regular hexagon, and so forth. When dealing in the third dimension, a regular solid is one in which all of the faces are congruent regular polygons and the angles between any two faces are the same. These regular solids are called **polyhedrons.** A good example, and the most familiar, is the cube. It is interesting to note that instead of an infinite number of regular polyhedrons (poly—many, hedron—faces), there are five, and only five, regular polyhedrons. They are:

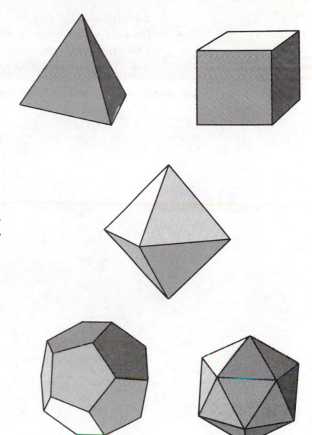

These are the only possible regular polyhedra: the tetrahedron, hexahedron (or cube), octahedron, dodecahedron, and the icosahedron.

With a little thought, perhaps we can see that we cannot form the corner of a polyhedron with less than three faces, and that a corner may be formed by joining three, four, or five equilateral triangles. Six equilateral triangles would flatten out to be a plane. Three squares might make a corner, but four would flatten out to a plane. For the same reasons only three pentagons can be used at a vertex. Hexagons and all other regular polygons would not work, as three or more could not form a corner. This argument for the possible figures that can be used to form regular polyhedrons was the source for Euler's Formula.

Let us put some of the information about the regular polyhedra in a table:

|  | # of Faces | Type | # of Edges | # of Vertices |
|---|---|---|---|---|
| Tetrahedron | 4 | triangle | 6 | 4 |
| Hexahedron | 6 | square | 12 | 8 |
| Octahedron | 8 | triangle | 12 | 6 |
| Dodecahedron | 12 | pentagon | 30 | 20 |
| Icosahedron | 20 | triangle | 30 | 12 |

Leonard Euler (1707–1783), one of the all-time great mathematicians, discovered an interesting formula that not only applies to the regular polyhedra, but applies to *any* polyhedron.

$$E + 2 = V + F$$

which says that in any polyhedron the number of edges is always 2 less than the sum of the number of vertices and the number of faces. Here are a few polyhedra to check it out on; you might find some other solids that have polygons for faces and see if this formula works for you.

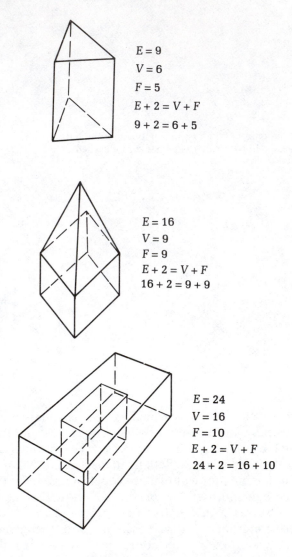

$E = 9$
$V = 6$
$F = 5$
$E + 2 = V + F$
$9 + 2 = 6 + 5$

$E = 16$
$V = 9$
$F = 9$
$E + 2 = V + F$
$16 + 2 = 9 + 9$

$E = 24$
$V = 16$
$F = 10$
$E + 2 = V + F$
$24 + 2 = 16 + 10$

In a given sphere, the inscribed dodecahedron is closer to the volume of the sphere than is the figure with more sides, the regular icosahedron. The cube comes closer to the volume of the sphere than does the octahedron. This ordering of volumes, and the proof that there exists only five regular polyhedrons is the final result of the thirteen books of Euclid.

Many natural crystals have polyhedral shapes.

The Greeks thought that the five regular polyhedra were the building blocks of matter, and in fact, some crystal structures do have these shapes. Some of this thought must have persisted until the time of Johann Kepler, the German astronomer and mathematician (1571–1630). Before he discovered and proved the three Kepler's Laws for which he is famous, he constructed a model of the solar system in which each planet was on a sphere and the spheres were inscribed in and circumscribed about the five regular polyhedra. The sun was in the center, of course. It was a very clever idea, and comes within 5 or 10 per cent of giving correct modern-day values.

For a reference on polyhedra that are not regular, methods of constructing them, and many other beautiful geometric patterns, see *Patterns In Space* by Col. R. S. Beard, published in 1973 by Creative Publications.

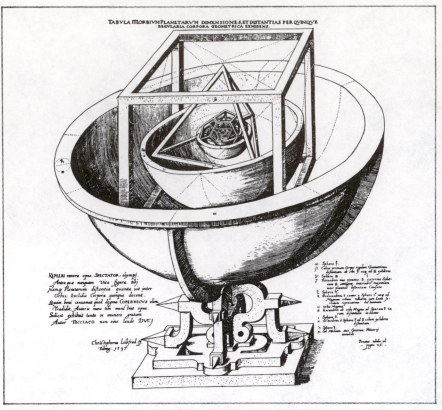

Kepler's model of the universe placed the orbits of the planets on spheres circumscribed about the five different regular polyhedra.

As you are no doubt aware, a floor can be tiled with equilateral triangles, squares, rectangles, hexagons, and many other shapes both regular and irregular. This is called the tessellation of a plane. Now the question arises: what polyhedra, regular or irregular, can be used to completely fill space? This is called the tessellation of space. Of course cubes, rectangular parallelepipeds, and tetrahedra will work—but what others? For a very fine discussion of tessellations of the plane and of space, along with a quite complete bibliography of further readings, you might refer to *Geometry: An Investigative Approach* by O'Daffer and Clemens, published in 1976 by Addison Wesley.

### A summary of volume formulas

| Volume of a | equals |
|---|---|
| rectangular parallelepiped | $lwh$ |
| cube | $a^3$ |
| any parallelepiped | $Bh$ |
| any prism | $Bh$ |
| any cylinder | $Bh$ |
| right circular cylinder | $\pi r^2 h$ |
| any pyramid | $\frac{1}{3}Bh$ |
| any cone | $\frac{1}{3}Bh$ |
| right circular cone | $\frac{1}{3}\pi r^2 h$ |
| sphere | $\frac{4}{3}\pi r^3$ |

1. How many cubic meters of air in a room measuring 8 m by 10 m by 3 m?

2. What is the volume of stone in an Egyptian pyramid (square base) 100 m on a side and having a height of $50\sqrt{2}$ for one of the triangular faces (slant height)?

3. Find the volume of a regular tetrahedron with a side of 12 cm.

4. Find the volume of a regular octahedron with an edge of 12 cm.

5. If an ice cream cone is completely filled with ice cream and has a hemisphere of ice cream on top, find the total volume of ice cream if the diameter of the top of the cone is 6 cm and the height of the cone (not including hemisphere) is 15 cm.

6. A semicircle of radius 20 cm is folded to form a cone. What is its volume?

7. A soda can has a diameter of 6.5 cm and a height of 12 cm. What is its volume?

8. A cylindrical water tank has a circumference of 90 ft. and a height of 27 ft. How many gallons of water does it hold? (Use 3.14 for $\pi$ and 7.48 gallons of water per cubic foot.)

9. A hexagonal prism with sides of 2 cm is 15 cm in height. What is its volume?

10. A spherical propane tank is 2 m in diameter. What is its volume?

11. Which contains the most tomatoes; a can of tomatoes with diameter of 8 cm and height of 12 cm, or 2 tomatoes (assume they are spherical) with a diameter of 8 cm?

12. An edge of the base of a regular right hexagonal prism is 4 cm and it has a height of 9 cm. Find the volume.

13. The base of a right parallelepiped is a rhombus whose diagonals are 10 cm and 24 cm. The altitude of the parallelepiped is 7 cm. Find its volume.

14. The lateral area of a right circular cylinder is 132 cm² and the circumference of its base is 12 cm. Find its volume.

15. The total area of a right circular cylinder is $56\pi$ cm² and the altitude is 12 cm. What is its volume?

16. The hypotenuse of an isosceles right triangle is 6 cm. Find the volume of the cone generated if the triangle is revolved about one of its legs.

17. The radius of the base of a right circular cylinder is 8 cm and the altitude is 15 cm. A right circular cone is placed inside the cylinder so that their bases coincide, and the vertex of the cone just touches the upper base of the cylinder. What is the ratio of their volumes?

18. A sphere is circumscribed about a cube with edge of 6 cm. Find the volume of the sphere.

19. Find the volume of a pyramid with a square base, where every edge of the pyramid is 4 cm.

20–24. Verify Euler's Formula for each of these polyhedra:

20.

21.

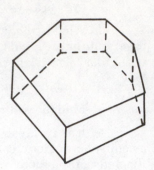

22.

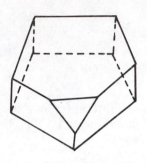

23.

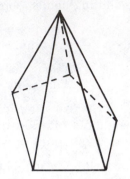

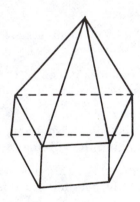

24.

25. Find the volume of a conical pile of topsoil, if the diameter of its base is 3 m and it forms an angle of 45° with the ground.

*26. Find the volume of a regular tetrahedron inscribed in a unit sphere.

27. A cake recipe is intended for an angel food cakepan that is a cylinder 4 in. high and 10 in. across the bottom with a 2 in. diameter hole in the middle. Will this recipe fit into an 8 × 8 × 4 in. glass baking pan?

28. A drinking cup in the shape of a frustum of a cone is to measure 7 cm across the top, 5 cm across the bottom, and hold 200 ml (about 6 oz.). How tall will it have to be?

29. How much water can be held by a trash barrel in the shape of a frustum of a cone with upper diameter of 20 in., bottom diameter of 16 in., and height of 25 in.? (1 cu. ft. ≈ 7.48 gal.)

30. How much will the cement cost to construct 12 hollow pillars with an outer diameter of 1 m and an inner diameter of $\frac{1}{2}$ m and height of 4 m if cement costs $45.00 per cubic meter?

31. A 3-4-5 right triangle is rotated about the 3-unit leg. What is the volume of the cone generated?

32. A 3-4-5- right triangle is rotated about its hypotenuse. What is the volume of the solid generated?

33. A feed-dispensing bin is in the shape of a cylinder on a cone. If the cone and the cylinder each have a height of 2 m, and the vertex of the cone is 60°, how many bushels will the bin hold? (1 bushel ≈ 35.238 liters)

34. If fill dirt costs \$2.90 per cu. m, how much will it cost to fill a hole 5 m deep, 12 m wide, and 17 m long?

35. A feed trough is in the shape of a trapezoidal prism. The trapezoid on the ends has upper base of 50 cm, lower base of 30 cm, and a height of 30 cm. The trough is 3 m long. How much feed will it hold when full?

36. A propane tank is in the shape of a cylinder with hemispheres on each end. The entire length of the tank is 3 m and its diameter is 1 m. What is its volume?

37. A 3 × 3 × 3 cube is painted green. It is then cut into 27 1 × 1 × 1 cubes. How many of these cubes have 3 green faces, how many have 2 green faces; one green face; none?

*38. How many saw cuts are necessary to cut the above 3 × 3 × 3 cube into the 27 small 1 × 1 × 1 cubes, if after each cut it is possible to rearrange and stack previously cut sections and cut several at a time?

39. A cube with an edge of 12 cm is cut into 6 congruent pyramids—each face of the cube is the base of one of the pyramids. Find the volume of one of these pyramids by using the formula for the volume of a pyramid, and then verify that it is one-sixth of the volume of the cube.

*40. A regular tetrahedron is one where all four faces are equilateral triangles. How many regular tetrahedrons with edge of 1 unit are necessary to form a regular tetrahedron with an edge of 2 units? with an edge of 3 units? with an edge of $n$ units?

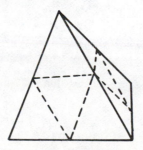

*41. Show, either with a proof, or by numerical example, that when the five regular polyhedra are inscribed in a unit sphere, the polyhedra with the greater number of vertices give the best approximations of the volume of the sphere. That is;

$$V_{\text{dodecahedron}} > V_{\text{icosahedron}} > V_{\text{cube}} > V_{\text{octahedron}} > V_{\text{tetrahedron}}$$

*42. Construct a model of each of the five regular polyhedra. A technique for constructing them with cardboard and rubber bands is shown in the center section of *The Mathematics Teacher*, April, 1977.

## 7–Summary

Definitions of some solids:

Rectangular parallelepiped—a solid with six rectangular faces

Cube—a solid with six square faces

Prism—the solid formed when the corresponding vertices of two congruent polygons, lying in parallel planes, are joined.

Cylinder—the solid formed when the corresponding points of two congruent circles, lying in parallel planes, are joined—forming a curved lateral surface.

Pyramid—the solid formed when each vertex of a polygon is joined to some point not in the plane of the polygon.

Cone—the solid formed when each point of a circle is joined to some point not in the plane of the circle—forming a curved lateral surface.

Sphere—the set of points a given distance (radius) from a given point (center).

Zone—the surface of a sphere cut off by two parallel planes.

Frustum of a cone—the portion of a cone cut off by two parallel planes.

Surface Formulas:

| | |
|---|---|
| Rectangular parallelepiped | $2(lh) + 2(wh) + 2(lw)$ |
| Cube | $6a^2$ |
| Prism (right or oblique) | $ph + 2B$ |
| Circular cylinder (right or oblique) | $2\pi r(h + r)$ |
| Pyramid (right regular) | $\frac{1}{2}pl + B$ |
| Right circular cone | $\pi r(l + r)$ |
| Frustum of a cone | $2\pi r_M l$ |
| Sphere | $4\pi r^2$ |

Volume Formulas:

| | |
|---|---|
| Rectangular parallelepiped | $lwh$ |
| Cube | $a^3$ |
| Any parallelepiped | $Bh$ |
| Any prism | $Bh$ |
| Any cylinder | $Bh$ |
| Right circular cylinder | $\pi r^2 h$ |
| Any pyramid | $\frac{1}{3}Bh$ |
| Any cone | $\frac{1}{3}Bh$ |
| Right circular cone | $\frac{1}{3}\pi r^2 h$ |
| Sphere | $\frac{4}{3}\pi r^3$ |

The Grainger Collection, New York.

# CHAPTER 8

# SUPPLEMENTARY TOPICS

In this chapter, we have some topics which can be omitted from the usual sequence without destroying the continuity of the material. Not all geometry texts include these topics, but the subject is much richer and more interesting if the time can be found to include them. However none of the sections 1–25 should be omitted in order to include these topics. They can be covered after the completion of the first 25 sections, or as indicated below.

In some of the previous theorems we made use of the Euclidean Parallel Postulate, which we gave as "Through a given point, only one line can be drawn parallel to a given line." Actually, Euclid stated it in a different, but equivalent, form: "If a transversal falling on two straight lines makes the interior angles on the same side of the transversal less than two right angles, the two straight lines, if produced indefinitely, will meet on that side of the transversal." This postulate is not only necessary for our work with the parallel lines, but is fundamental to our work with the measure of the angles of a triangle, similarity, quadrilaterals, and many other basic concepts.

Since a deductive science should have as few axioms and postulates as possible, it would be nice if a postulate as obvious as Euclid's Parallel Postulate could be proven, in terms of other postulates and axioms, and could then be a theorem. Mathematicians tried for 2000 years without success to do this. In the nineteenth century, it was finally an indirect approach that yielded results. Since there are only three possible cases: exactly one line through a point parallel to a given line, no lines through the point parallel to a given line, or more than one line through the point

## 26. Non-Euclidean geometries

and parallel to a given line—all we have to do is assume, in turn, the second and third possibilities, reduce them to an absurdity, and then we will have proven the first of the three possibilities is true. But the results are surprising. Mathematicians discovered there are no contradictions when either of the second or third possibilities replace Euclid's postulate!

This lack of contradictions tells us that the Parallel Postulate is independent of the other postulates, and as such can not be proven from them. When a different postulate is used in its place, many of the theorems turn out differently, but there are no contradictions among them. The resulting geometry is called a non-Euclidean geometry.

Two different men developed independently, at about the same time, the geometry assuming that there was more than one line through a point and parallel to a given line. John Bolyai (1802–1860)—a Hungarian army officer and Nicholas Lobachevski (1793–1856)—a Russian professor of mathematics. Lobachevski was the first to publish, so the geometry often bears his name.

It should be noted that in applications it is impossible to differentiate between any of the three geometries. In the Lobachevskian geometry, the sum of the angles of a triangle will be less than 180°, exactly 180° in Euclidean, and more than 180° in the third type (Riemannian). The small distances with which we work (such as from here to a star) are too small a part of the total universe to determine by measurement which is the actual case. This is similar to a surveyor who does not have any noticeable error if he fails to take into consideration that he is on the surface of a sphere, and not on a plane.

It may seem difficult to imagine how there could be more than a single line through a point parallel to a given line. But remember, our definition did not really give us a way to test whether two non-intersecting lines were actually parallel. Is our universe bounded or unbounded? Does it have an "edge" that we are simply unaware of? If we live in a bounded universe, a superbeing, able to see the entire picture, would find it quite reasonable to have many lines through point $P$ parallel to line $l$. That is, they do not intersect $l$ even when extended to the ends of the universe!

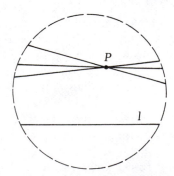

Bernhard Reimann (1826–1866)—a German mathematician, investigated what would happen if Euclid's Parallel Postulate were replaced by one saying there were no parallels through a point $P$. Again there were no inconsistencies. The resulting geometry is called Riemannian geometry. In it the sum of the angles of a triangle is greater than 180°. A good model for this geometry is the surface of a sphere. A triangle formed by two lines of longitude and the equator will contain two right angles.

Also, any two lines of longitude will both be perpendicular to the equator but will not be parallel—they will intersect at the poles. This differs from our Euclidean theorem that two lines perpendicular to the same line are parallel.

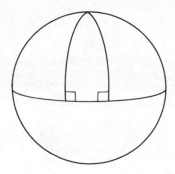

Notice that for a small area on the surface of a large sphere, the surface is so nearly "flat" that Euclidean geometry seems to describe everything adequately. It is only when we travel further, or get a larger view of things, that spherical geometry becomes necessary and what was seen as "flat" is actually "curved". Likewise, our view of the universe as Euclidean may be valid only in a smaller sense and further space travel may reveal the need for a different geometry.

Until fairly recently, the geometry of Euclid was considered as the true description of the world around us, and the non-Euclidean geometries were mathematical curiosities. In fact, the idea that Riemannian or Lobachevskian geometry might give a better description of space is still difficult to accept. However, Einstein used Riemannian geometry in his general theory of relativity—and currently, the best bet is that it gives a better description of space and the universe than any of the other geometries. Of course, in small scale practical applications, the three are indistinguishable.*

**EXERCISE SET 26**

1. Gauss attempted to determine by experimentation which of the three geometries described the universe. He put three observers on three mountaintops in Germany, and then very carefully measured the angles between them. If the sum of the angles was *exactly* 180°, the universe would be Euclidean; less than 180°, then it would be Lobachevskian; and a sum greater than 180° would make it Riemannian. Explain why this experiment was doomed to failure from the start. (Incidentally, the triangle measured a few seconds less than 180°.)

2. On the surface of a sphere, let "line" mean the shortest distance between two points; that is, a "great circle," which can be found on a globe by stretching a string between the two points. Let "parallel lines" be lines that do not intersect. Through a given point, say Topeka, how many lines can be drawn parallel to the equator? Which geometry does this represent?

*For good discussion of the curvature of space see: "The Curvature of Space in a Finite Universe" by J. J. Callahan, in *Scientific American*, August, 1976, pp. 90–100, and "Large and Small; Exploring the Laws of Nature" by E. Creutz, in *Science Teacher;* September, 1976; pp. 27–31.

3. Suppose the universe is finite and has an "edge." How many lines through P can be drawn parallel to L? What geometry does this represent?

4. Three lines (great circles) on the surface of a sphere intersect to form a triangle. They are the equator, and the lines of longitude 20°W and 110°W. What is the sum of the angles of the triangle formed? Which geometry is this?

5. Soap films always take the shape which has the least area. The shape a film would take between two loops is called a hyperboloid.

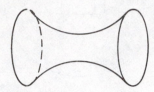

If a triangle is drawn on this surface what will be the sum of the angles? Which geometry does this represent? (If you need a physical object to work with, a vase or a trumpet might be the shape of the right or left half, and a saddle is the shape of the top half. A "line" is still defined as the shortest distance between two points.)

6. On the hyperboloid, how many lines can be drawn through a point P, which are "parallel" to a line L? Which geometry is this?

7. Can a quadrilateral be drawn on the surface of a sphere in such a manner that the angles are all right angles? Explain.

8. There are not supposed to be any "parallel lines" on the surface of a sphere. Why does this figure not represent parallel lines?

*9. Prove that in Riemannian Geometry, an angle inscribed in a semicircle is greater than a right angle.

*10. Prove that in Lobachevskian Geometry an angle inscribed in a semicircle is less than a right angle.
Suggested figure for 9 and 10:

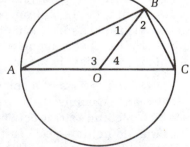

Prove $\angle ABC \lesseqgtr 90°$

*11. Report on the article: "The Curvature of Space in a Finite Universe", by J. J. Callahan in *Scientific American*, August, 1976.

*12. Research, and write a biography of Riemann, of Lobachevski, or of both.

*Photograph by Harold M. Lambert.*

If "converse" is what we call prison poetry, then perhaps "concurrency" is a name for counterfeit money. But we should probably look for a geometrical definition of concurrency. **Two or more lines which share a common point** are said to be **concurrent.**

Any two non-parallel lines will be concurrent in some point. What we wish to examine here is when three lines will happen to be concurrent. Three lines might have a number of different configurations:

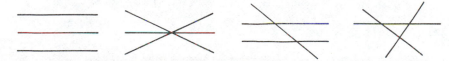

They may have 0, 1, 2, or 3 points of intersection. (How many points of intersection are possible with four lines?) We wish to look at some situations where three lines have only one point of intersection, that is, they are concurrent.

In *any* triangle:

> the angle bisectors are concurrent,
> the perpendicular bisectors of the sides are concurrent,
> the altitudes are concurrent, and
> the medians are concurrent.

Recall that an altitude is a line from a vertex, perpendicular to the opposite side or to the opposite side extended. It is in the case of an obtuse triangle that the opposite side must sometimes be extended in order to draw the perpendicular.

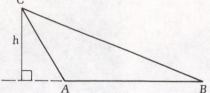

A **median** of a triangle is a line from a vertex to the midpoint of the opposite side. *AD* is a median if *BD = CD*.

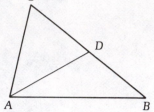

Let us examine some cases for the three main types of triangles.

For the angle bisectors:

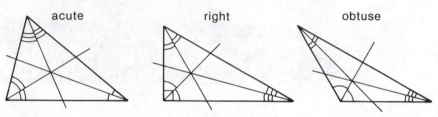

The angle bisectors are concurrent in a point which is called the **incenter.** It is the center of the inscribed circle for the triangle. To inscribe a circle in a triangle, simply construct the bisectors of two of the angles (the third one is not necessary as two lines determine a point) and their intersection is the incenter. The perpendicular distance from the incenter to any side is the radius.

For the perpendicular bisectors of the sides:

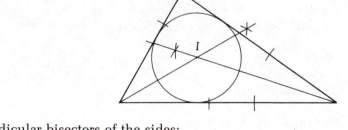

The perpendicular bisectors of the sides are concurrent in a point called the **circumcenter.** It is the center of the circumscribed circle. To draw a circle around a triangle, simply construct the perpendicular bisectors of two sides (again, the third line is not needed, as two will determine the point) and that will locate the circumcenter. Use the distance from the circumcenter to any one of the vertices as the radius and draw the circle.

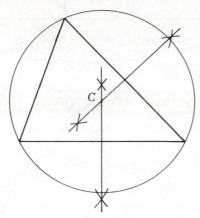

For the altitudes:

acute right obtuse

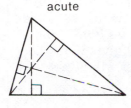 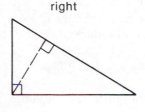 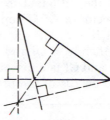

The altitudes are concurrent in a point called the **orthocenter.** The orthocenter is pretty worthless as far as practical uses are concerned. It is used in other theorems and relationships, such as the 9-point circle discussed later.

For the medians:

acute right obtuse

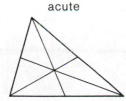 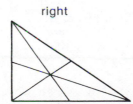 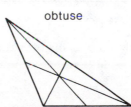

The medians are concurrent in a point called the **centroid,** or the center of gravity. In a practical sense it is the most useful of all the points of concurrency. In building construction, if we wish to support a triangular steel plate, the support should be placed at the centroid. It is the point where all of the weight is centered. If a triangular piece is to be hung in a mobile, the string should be attached at the centroid. The triangular piece will then hang in a horizontal position. If a non-triangular shape is to be supported or balanced, sometimes it can be divided into a number of triangles, the centroid of each triangle located, and then these centroids "averaged" (a process we will not discuss) to find the centroid for the entire piece.

For each of the four points of concurrency we need to *prove* that in *all cases*, not just the examples we drew, that the specified lines will be concurrent.

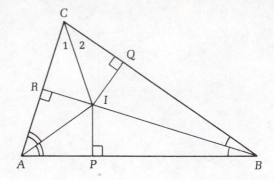

| The angle bisectors of a triangle are concurrent. |
|---|

*Given:* $AI$ bisects $\angle A$, $BI$ bisects $\angle B$

*Prove:* $CI$ bisects $\angle C$

(Analysis: Since the bisectors of $\angle A$ and $\angle B$ will meet at some point, $I$, it will be sufficient to show concurrency if we can show that $CI$ is the bisector of $\angle C$. Drop perpendiculars to each of the sides from $I$. Show $\triangle AIR \cong \triangle AIP$ giving $RI = PI$. Show $\triangle BIP \cong \triangle BIQ$ giving $PI = QI$. So then $RI = QI$ and we can show $\triangle RIC \cong \triangle QIC$. This will give $\angle 1 = \angle 2$ and $CI$ is a bisector.)

The details of the proof are left to the student.

| The perpendicular bisectors of the sides of a triangle are concurrent. |
|---|

*Given:* $PQ \perp$ bis. of $AB$, $PR \perp$ bis. of $BC$

*Prove:* $PS \perp$ bis of $AC$

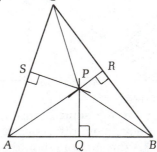

(Analysis: $QP$ and $RP$ will meet at *some* point $P$ since perpendiculars to non-parallel lines will intersect. Concurrency of the perpendicular bisectors will be demonstrated if we can show that the line $SP$ is the perpendicular bisector of $AC$. Draw lines from $P$ to each vertex. Show that $\triangle APQ \cong \triangle BPQ$ giving $AP = BP$. Show $\triangle BRP \cong \triangle CRP$ giving $BP = CP$. This will give $AP = CP$ and we can show $\triangle ASP \cong \triangle CSP$. This will allow us to show $PS$ is the perpendicular bisector of $AC$.)

The details of the proof are left to the student.

> ### *The altitudes of a triangle are concurrent.*

Given: In $\triangle ABC$ $AQ \perp BC$, $BR \perp AC$, $CP \perp AB$

Prove: $AQ$, $BR$, $CP$ are concurrent

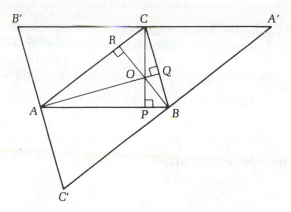

| | |
|---|---|
| 1. Draw $A'B'$ through $C$ parallel to $AB$ <br> $B'C'$ through $A$ parallel to $BC$ <br> $A'C'$ through $B$ parallel to $AC$ | 1. Construction |
| 2. $AQ \perp BC$, $BR \perp AC$, $CP \perp AB$ | 2. Given |
| 3. $AQ \perp B'C'$, $BR \perp A'C'$, $CP \perp A'B'$ | 3. Alt. int. angles |
| 4. $ABCB'$ is a parallelogram | 4. Opp. sides parallel |
| 5. $AB = B'C$ | 5. Opp. sides of $\square$ are $=$ |
| 6. $ABA'C$ is a parallelogram | 6. Opp. sides parallel |
| 7. $AB = A'C$ | 7. Opp. sides of $\square$ are $=$ |
| 8. $B'C = A'C$ | 8. Substitution |

In like manner $C'B = A'B$ and $B'A = C'A$

| | |
|---|---|
| 9. $AQ \perp$ bis. of $B'C'$, $BR \perp$ bis. of $A'C'$, $CP \perp$ bis. of $A'B'$ | 9. Def. of $\perp$ bis. |
| 10. $AQ$, $BR$, $CP$ are concurrent | 10. $\perp$ bis. of $\triangle A'B'C'$ are concurrent |

In the following theorem, we not only show that the medians are concurrent, but also show where the point of concurrency will be.

> *The medians of a triangle are concurrent in a point $\frac{2}{3}$ of the distance from any vertex to the opposite side.*

**NOTE:** Don't be discouraged if you find this proof difficult to follow. It is very long and involved. It would really be oppressive if every step were given. We are not so much concerned with the details of this proof as we are with the deductive reasoning process involved and the resulting relationship.

Given: *AE, BF, CD* are medians

Prove: *AE, BF, CD* are concurrent
$AG = \frac{2}{3}AE$ and $BG = \frac{2}{3}BF$
$CG = \frac{2}{3}CD$

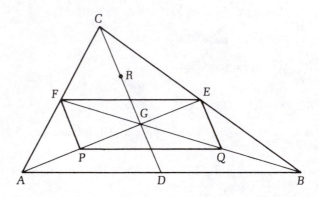

(Analysis: By using similar triangles show that the opposite sides of *PQEF* are equal and it is therefore a parallelogram. Since the diagonals of a parallelogram bisect each other, *PG = GE* and since by construction *PG = AP*, then $AG = \frac{2}{3}AE$. Show the concurrency by showing median *CD* must go through *G*, the intersection of *AE* and *BF*.)

| | |
|---|---|
| **1.** Bisect *AG* at *P* and *BG* at *Q* | **1.** Construction |
| **2.** Draw *PQ, QE, EF, FP* | **2.** Construction |
| **3.** *AE, BF, CD* are medians | **3.** Given |
| **4.** *AF = CF, BE = CE* | **4.** Def. of medians |
| **5.** $\dfrac{AF}{CF} = \dfrac{BE}{CE} = 1$ | **5.** Division |
| **6.** $\dfrac{AC}{CF} = \dfrac{BC}{CE} = \dfrac{2}{1}$ | **6.** Proportional addition |
| **7.** $\angle C = \angle C$ | **7.** Reflexive |

| | |
|---|---|
| **8.** $\triangle CEF \sim \triangle ABC$ | **8.** SAS, sim. |
| **9.** $\dfrac{AB}{FE} = \dfrac{2}{1}$ | **9.** Corr. sides prop. |

In like manner, using $\triangle PQG \sim \triangle ABG$ show $\dfrac{AB}{PQ} = \dfrac{2}{1}$

| | |
|---|---|
| **10.** $PQ = FE$ | **10.** 3 terms of a prop. $=$, 4th are $=$ |

Now use the same procedure to show $EQ = PF$

| | |
|---|---|
| **11.** $PQEF$ is a parallelogram | **11.** Opp. sides are $=$ |
| **12.** $PG = GE$ and $QG = GF$ | **12.** Diagonals bisect ea. other |
| **13.** $PG = AP$ and $QG = BQ$ | **13.** Def. of bisect |
| **14.** $AP = PG = GE$ and $BQ = QG = GF$ | **14.** Substitution |
| **15.** $AG = \frac{2}{3}AE$ and $BG = \frac{2}{3}BF$ | **15.** Substitution (algebraic) |

This shows the desired numerical relation. We now show concurrency. $AE$ and $CD$ meet at some point $G'$ (not shown on the figure because we are going to show coincidence, and the figure is messed up enough already.) By bisecting $CG'$ at $R$ and drawing $PDER$ we can use the above method to show $AG' = \frac{2}{3}AE$ and $CG' = \frac{2}{3}CD$. But since $AG = \frac{2}{3}AE$

| | |
|---|---|
| **16.** $AG = AG'$ | **16.** Subsittution |
| **17.** $\therefore G'$ coincides with $G$ | **17.** Dist. Postulate |
| **18.** $AE, BF, CD$ are concurrent | **18.** They all pass through $G$ |

Now, let us see if we can make use of the preceding theorems in some examples.

**An equilateral triangle has sides of 2 cm. Find the radius of the circumscribed circle and the radius of the inscribed circle.**

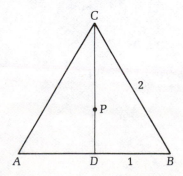

The median, $CD$, is also the perpendicular bisector of the side and the bisector of the angle. So $P$ is the circumcenter, incenter, and the centroid. Because $P$ is a centroid, $CP = \frac{2}{3}CD = \frac{2}{3}\sqrt{3}$ and $PD = \frac{1}{3}CD = \frac{\sqrt{3}}{3}$. $CP$ is the radius of the circumscribed circle and $PD$ is the radius of the inscribed circle.

**The roof of an A-frame has the shape of an isosceles triangle with a pitch of 60°. How far from the peak is the center of gravity of a cross section? The edge of the roof measures 6 m.**

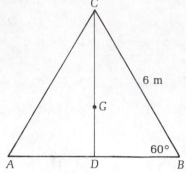

Since median $CD$ is also perpendicular to $AB$ we have a 30-60-90 triangle. If $CB = 6$, $DB = 3$, and $CD = 3\sqrt{3}$. The centroid, $G$, is $\frac{2}{3}CD$ or $2\sqrt{3}$ m from the peak.

**The medians of a triangle are 12, 15, and 20. How far from the vertices is the centroid?**

The distance from a vertex to the centroid is $\frac{2}{3}$ of the median.

$$\tfrac{2}{3}(12) = 8, \ \tfrac{2}{3}(15) = 10, \ \tfrac{2}{3}(20) = 13\tfrac{1}{3}$$

**A steel platform for a flagpole sitter is to be in the shape of an isosceles triangle with legs of 2 m and a base of 3 m. What point of the triangle should be attached to the pole for best balance?**

We wish to find the centroid—or center of gravity. It is at the intersection of the medians, which is $\frac{2}{3}$ of the length of any median from its vertex.

$$AC = 3, \ CD = \tfrac{3}{2}, \ BD \text{ is } \perp \text{ bis. of } AC$$

$$BD^2 = 2^2 - (\tfrac{3}{2})^2 = 4 - \tfrac{9}{4} = \tfrac{7}{4}$$

$$BD = \frac{\sqrt{7}}{2}$$

Therefore, the centroid is at a point $\frac{2}{3}\left(\frac{\sqrt{7}}{2}\right) = \frac{\sqrt{7}}{3}$ m from the vertex of the angle between the equal legs.

One additional topic of interest with regard to these points of concurrency is the "nine-point circle." It is a very interesting fact that certain points related to a triangle all fall on the same circle. We will not attempt to prove this theorem, but we will state it and illustrate it. You might find it worthwhile to try your own construction of a nine-point circle.

---

*In any triangle the feet of the three altitudes, the midpoints of the three sides, and the midpoints of the segments from the three vertices to the orthocenter, all lie on the same circle. This circle has its center at the midpoint of the line joining the orthocenter and the circumcenter.*

---

In the figure, $M_1$, $M_2$, $M_3$ are the midpoints of the sides
        $H_1$, $H_2$, $H_3$ are the feet of the altitudes
        $A_1, A_2, A_3$ are the midpoints of the lines from the vertices
          to the orthocenter
        $O$ is the orthocenter
        $C$ is the circumcenter
        $P$ is the center of the nine-point circle

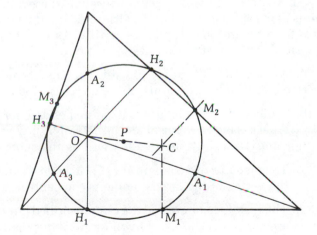

1. The medians of a triangle are 6 cm, 9 cm, and 10 cm. Find the lengths into which each is divided by their point of concurrency.

2. The equal legs of an isosceles triangle are each 8 cm long. The base angles are each 30°. How far from the vertex of the largest angle of the triangle do the medians intersect?

3. The centroid of an equilateral triangle is 4 cm from a vertex. Find the perimeter of the triangle.

4. An equilateral triangle has sides of 10 cm. What is the radius of the circumscribed circle?

5. Construct a triangle with sides of 6, 8, and 11 cm. Construct its inscribed circle.

6. Construct a circumscribed circle about a triangle with sides of 4, 6, and 9 cm.

7. Some points of concurrency are inside their triangles, some are on the sides of the triangle, and some are outside the triangle. Complete the following chart by locating the point of concurrency as "in," "on," or "out" of the triangle.

|  | incenter | orthocenter | centroid | circumcenter |
|---|---|---|---|---|
| **acute triangle** |  |  |  |  |
| **right triangle** |  |  |  |  |
| **obtuse triangle** |  |  |  |  |

8. There are many interesting theorems about the "nine-point circle". This is a circle which passes through these nine points: the midpoints of the sides, the feet of the altitudes, and the midpoints of the segments from the orthocenter to the vertices. Draw a triangle (a large one is easiest), and construct its nine-point circle.

9. An equilateral triangle has sides of 5 cm. Find the radius of the inscribed circle and the radius of the circumscribed circle.

10. A piece of colored glass is to be suspended horizontally in a mobile. The glass is in the shape of an equilateral triangle and measures 15 cm on a side. How far from a vertex will be the center of gravity?

11. The medians of a triangle are 15, 18, and 24 cm. How far from the vertex of the largest angle is the centroid?

12. The inscribed circle and the circumscribed circle of a triangle are concentric. What do you know about the triangle?

13. What is the radius of a circle circumscribed about a 3-4-5 right triangle?

14. What is the radius of a circle circumscribed about an isosceles right triangle with a leg of 6 cm?

15. What is the radius of a circle inscribed in an isosceles right triangle with a leg of 6 cm?

16. A 10 m (diameter) circular pool is 2 m from each of two fences of a rectangular lot. The owner wishes to run another fence diagonally as shown, 2 m from the pool and intersecting the present fences at equal

distances from the corner of the lot. How long will the new fence need to be?

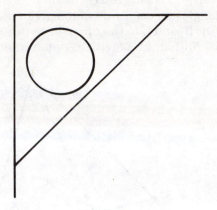

17. How large a circle can be inscribed in an isosceles right triangle with hypotenuse of 15 cm?

18. How small a circle can be circumscribed about an isosceles right triangle with legs of 21 cm?

*19. Prove: In an acute triangle $ABC$, with the feet of the three altitudes $D$, $E$, and $F$, the orthocenter of $\triangle ABC$ is also the incenter of $\triangle DEF$.

*20. Suppose three stores, $S_1$, $S_2$, and $S_3$ are located as shown and we wish to place a warehouse at some point $W$ so that the sum of the distances to the stores is a minimum. Where do we locate the warehouse?

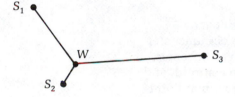

The solution to this problem of minimal distance is another point of concurrency for a triangle. It is called the Fermat Point (named after its discoverer, Pierre Fermat, 1601–1665). It is found by constructing equilateral triangles on each side of the given triangle and then joining their vertices, $A$, $B$, and $C$ to the opposite vertex of the original triangle. The lines $AS_1$, $BS_2$, and $CS_3$ will be concurrent, and that point is the Fermat Point.

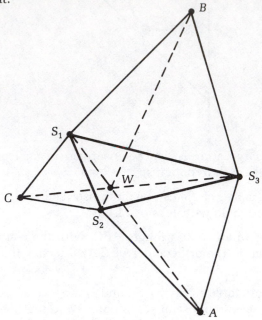

Construct a triangle, $PQR$, (for convenience, make it fairly large) and locate:

> $I$ - the incenter,
> $C$ - the circumcenter,
> $O$ - the orthocenter,
> $M$ - the centroid, and
> $F$ - the Fermat Point

Measure carefully the distances from each of these points to each vertex $P$, $Q$, and $R$. Show that the sum of $FP + FQ + FR$ is less than the sum of the distances from any of the other points to the three vertices.

**\*\*21.** Three spheres with radii of 13 are drawn with each of the vertices of triangle $ABC$ as their centers.

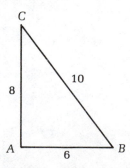

Where, above the plane of $\triangle ABC$, will be the common point of intersection of the three spheres? That is, where is the point that is 13 units from each vertex?

*22. Given any acute triangle *ABC*, how do we go about inscribing in it the triangle with the least perimeter? That is, *a* + *b* + *c* is a minimum.

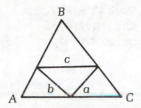

Draw a triangle *ABC* and inscribe two or three triangles, and by careful measurement find their perimeters. Now construct the three altitudes of △*ABC* and join the feet of those altitudes to form an inscribed triangle. This triangle is called the Orthic Triangle, and it is the inscribed triangle with minimum perimeter. Measure its perimeter and verify that it is less than that of the other triangles you tried.

**23. The previous problem, finding the inscribed triangle with minimum perimeter, is called Fagnano's Problem—named after the man who proposed it in 1775. Fagnano solved the problem by using calculus. We can prove that the Orthic Triangle has minimum perimeter with the use of some transformational geometry. Draw an acute triangle, along with its Orthic Triangle, and any other inscribed triangle. Now construct five successive reflections of the triangle and its contents (start at one end of a long paper!) Reflect about *AC*, *BC*, *AB*, *AC*, and then *BC*. Now compare the path formed by *a* + *c* + *b* + *a* + *c* + *b* (which is two perimeters of the Orthic Triangle) to the path formed by the successive sides of the other inscribed triangle (which will be twice its perimeter.) Since the path from *P* to its final reflection point *P'* is shorter than the path *QQ'*, it should be obvious that the perimeter of the Orthic Triangle is smaller.

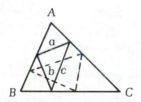

*24. Construct the altitude to side *AB* of triangle *ABC*, when the vertex, *C*, of the triangle has been cut off.

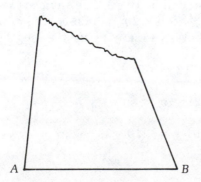

FRANK AND ERNEST                                    by Bob Thaves

Reprinted by permission of NEA.

## 28. Composite figures

With all the formulas you now have at your command, you can find perimeters, areas, and other parts of composite figures. By composite figures, we mean those which are composed of parts of other general types for which we have developed formulas. For example, consider this problem:

**We wish to find the length of a pulley belt which is wrapped around two wheels. The wheels have radii of 12 cm and 2 cm, and their centers are 20 cm apart.**

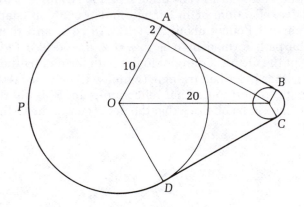

By adding a few auxiliary lines, we can see that the length of the belt is composed of four parts; two common external tangents, minor arc $\overset{\frown}{BC}$, and major arc $\overset{\frown}{APD}$. Since radii are perpendicular to the tangents at their points of contact, we have a right triangle with hypotenuse of 20 and one leg of 10. This will be a 30-60-90 triangle and the other leg, which is also equal to the tangent, will be $10\sqrt{3}$. The angle at $O$ will be the 60° angle, so $\overset{\frown}{APD} = 240°$. $\overset{\frown}{BC}$ will be 120°, which is 360° minus two right angles and two 30° angles. We can now figure the lengths of the arcs.

$$\overset{\frown}{APD} = \tfrac{240}{360}\, 2\pi 12 = 16\pi$$

$$\overset{\frown}{BC} = \tfrac{120}{360}\, 2\pi 2 = \tfrac{4}{3}\pi$$

Length of pulley = $2(10\sqrt{3}) + 16\pi + \tfrac{4}{3}\pi$ or $20\sqrt{3} + \tfrac{52}{3}\pi$ cm

As another example, consider a lot at the end of a cul-de-sac. The cul-de-sac has a radius of 15 m, the side of the lot is 45 m, and it is 60 m across the back of the lot. If the side lines are extended to the center of the cul-de-sac, an equilateral triangle is formed. Find the area of the lot.

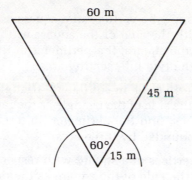

The area of the lot can be determined by subtracting the area of a sector from the area of a triangle.

$$A = \frac{\sqrt{3}}{4}s^2 - \frac{60}{360}\pi r^2$$

$$A = \frac{\sqrt{3}}{4}60^2 - \frac{15^2}{6}\pi = 900\sqrt{3} - \frac{75}{2}\pi \text{ m}^2$$

To the nearest m², about 1441 m².

Some composite problems, when properly analyzed, can even be solved mentally.

**Find the perimeter of the quatrefoil inscribed in a square with side 10.**

If one sees that the figure is composed of four semicircles, which would be two circles with diameter 10, then the perimeter is $2(\pi d) = 2(10\pi) = 20\pi$.

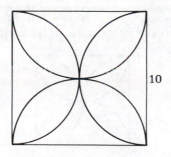

*Perimeters of composite figures—leave answers in exact form.*

1. With each side of an equilateral triangle as a diameter, semicircles are drawn dividing the triangle into four parts. If the side of the triangle is 6 cm, find the perimeter of each part.

2. With two opposite vertices of a square as respective centers, circles are drawn with the side of the square as their radius. Find, in terms of the side of the square, the perimeter of the figure within the square bounded by the two intersecting arcs.

3. With the vertices of an equilateral triangle as centers, circles are drawn with half the side of the triangle as the radius of each. If each side of the triangle is s, find the perimeter, in terms of s, of the figure within the triangle bounded by three arcs.

4. With the vertices of a square with respective centers, circles are drawn with half the side of the square as the radius of each. If the side of the square is s, find the perimeter of the figure within the square bounded by the four arcs.

5. Three equal circular plates with radius of 5 cm are placed such that each is tangent to the other two. Find the length of a chord drawn taut about the three plates.

6. A belt is passed around two pulley wheels whose radii are 2 cm and 8 cm respectively, and whose line of centers is 12 cm. Find the length of the belt.

7. If a side of a given equilateral triangle is 3 cm, find the outside perimeter of the trefoil formed by arcs of three equal circles drawn with the vertices of the given equilateral triangle as respective centers and with radii equal to half the sides of the given triangle.

8. If a side of a given square is 4 cm, find the perimeter of the quatrefoil formed by arcs of four equal circles with the sides of the square as respective diameters.

9. The diameter, *AB*, of a circle is trisected and semicircles are drawn as shown in the figure. Find the perimeter of the shaded portion if *AB* is 9 cm.

10. A path 3 ft. wide goes around a rectangular garden measuring 10 ft. by 15 ft. How many petunias are needed to line both sides of the path, if they are planted 8 in. apart?

11. A flower bed in the shape of a square 2 m on a side, with semicircles on each side, is to be bordered with cement mowing strips which cost $1.25 a meter. What will be the total cost?

12. A large commercial building, with the dimensions shown, is to have a raingutter around its entire perimeter. How much raingutter is needed?

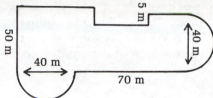

13. To minimize wear, the edge of a cam is to be equipped with a special wear-resistant strip. The strip costs 83¢ a centimeter. What will be the cost?

14. How long is the pulley belt around these two flywheels?

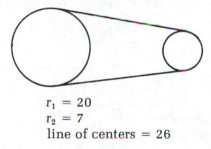

$r_1 = 20$
$r_2 = 7$
line of centers $= 26$

15. What is the perimeter of Ying (the dark part) in this Ying-Yang symbol?

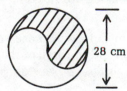

28 cm

*Areas of composite figures*—Leave answers in exact form.

16. In the adjacent figure, $\triangle ABC$ is equilateral with each side 24 cm. With $A$ as a center and with a radius equal to $\frac{1}{2}AB$ an arc is drawn cutting two sides of the triangle. Arcs are similarly drawn with $B$ and $C$ as centers. Find the area enclosed by the three arcs.

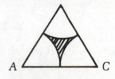

17. The "salt celler" curve is formed by the semicircle on $AB$ as a diameter, and the three smaller equal semicircles as shown in the figure, with the diameter of each being one-third of $AB$. If $AB = 12$ cm, find the area enclosed by the four semicircles.

18. The outline of the design of the Rainbow Division, World War I, is shown in the figure. $AB$, the diameter of the circle is trisected at $D$ and $E$, and semicircles are drawn as shown with radii of one-third and two-thirds of $AB$. Find the area of the shaded portion, if $AB = 6$ cm.

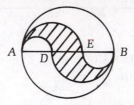

19. $ABCD$ is a square, $E$ bisects $AD$, $G$ bisects $BC$. With $G$ and $E$ as respective centers, arcs $\overparen{CF}$ and $\overparen{AF}$ are drawn. If the side of the square is 5 cm, find the area of the shaded portion.

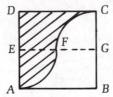

20. Find the area of a quatrefoil formed by semicircles whose respective diameters are the sides of a square with a diagonal of $4\sqrt{2}$.

21. A circle has a radius of 12 cm. With any point on the circle as center, and with the same radius, a second circle is drawn. Find the area common to both circles.

22. The radii of two circles are 7 cm and 1 cm, and their line of centers is 12 cm. A belt is passed around the two circles and drawn taut. Find the area enclosed.

23. An isosceles trapezoid with bases of 2 cm and 6 cm is circumscribed about a circle. Find the area of that portion of the trapezoid which lies outside of the circle.

24. With the vertices of an equilateral triangle as respective centers and with a side of the triangle as the radius, arcs are drawn so that each intersects the other two. If the side of the triangle is 2 cm, find the area enclosed by the figure.

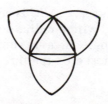

25. Find the area of Ying in the Ying-Yang symbol of problem 15.

26. Three circular steel plates of radii 18 cm are tied together with cord. What is the area included by the cord?

27. What length of 50 cm wide wallpaper is required to cover this kitchen wall? (Make the unrealistic assumption that there is no waste due to difficulty in matching patterns.)

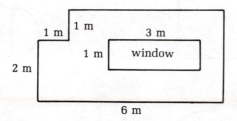

28. A lawn with the illustrated shape needs fertilizer. The recommended amount is 2 lbs. per 100 sq. ft. How much should be purchased?

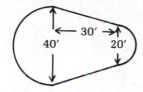

29. A horse on a 12 m rope is tied to the outside corner of an 18 × 30 m barn. Over how much area can he graze?

30. How much area can he graze, if tied to the middle of the short side of the barn?

31. How much area can he graze, if tied on the short side 10 m from the corner?

*32. Over how much area can a horse graze, if tied with a 20 m rope to the outside of a 20 m (radius) training ring?

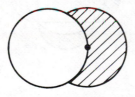

33–39. Find the following areas:

33.

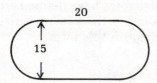

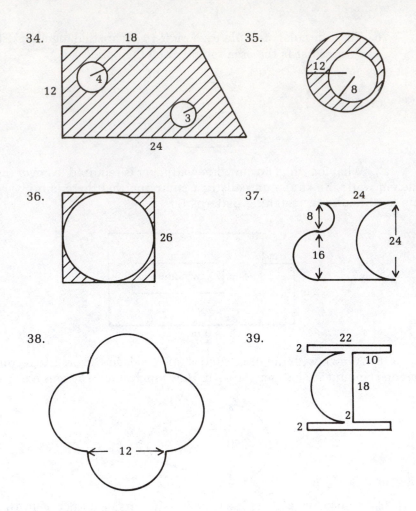

*40. Find the ratio of the area of the four lunes to the area of the square. (A **lune** is formed by a semicircle and an arc.)

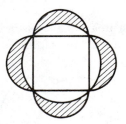

**29. The
Pythagorean
Theorem as
area**

The method of proving the Pythagorean Theorem we used earlier was not the original approach. In the original, it was the square on the hypotenuse, and in our version we handled it numerically as the square *of* the hypotenuse. Now that we have defined area and know some area formulas, we can consider this theorem as Pythagoras probably proved it. It is much more difficult in this form.

> *In right-angled triangles the square on the side subtending the
> right angle is equal to the sum of the squares on the sides con-
> taining the right angle.*

Given: $\triangle ABC$ with right angle at $A$

Prove: $A_{BCDE} = A_{ABFG} + A_{ACHK}$

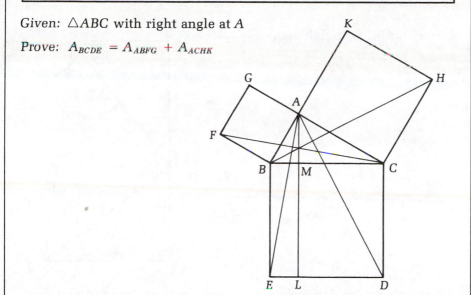

Since we have proved this theorem in another form previously, we will
only outline this form of the proof. The student should be able to fill in
the details.

Draw $AL \perp DE$
Draw $AE, BH, CF$
Show by addition of equals that $\measuredangle ABE = \measuredangle FBC$
Show by SAS that $\triangle ABE \cong \triangle FBC$
Show $\triangle ABE$ has same base and altitude as $BELM$
Show $\triangle FBC$ has same base and altitude as $ABFG$
So by doubles of equals $A_{BELM} = A_{ABFG}$
In like manner, using triangles $ACD$ and $BCH$ show that $A_{CDLM} = A_{ACHK}$
Then use addition to obtain the desired result:
$A_{BCDE} = A_{ABFG} + A_{ACHK}$

There are many, many very clever proofs of the Pythagorean Theorem that use the concept of dissecting the areas of the squares and putting them together again to form new squares. Many other proofs use algebraic relationships to prove $a^2 + b^2 = c^2$. James Garfield, the 20th president of the United States, is credited with this proof:

Given: $\triangle ABC$ with right angle C

Prove: $a^2 + b^2 = c^2$

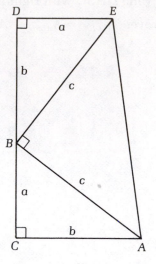

Again, we will only outline the proof and leave the details to the student.

Copy $\triangle ABC$ in the position shown as $\triangle BDE$
Draw $AE$ to form a trapezoid
$A_{ACDE} = A_{ABE} + 2A_{ABC}$
$\frac{1}{2}(a + b)(a + b) = \frac{1}{2}c^2 + 2(\frac{1}{2}ab)$
$(a + b)^2 = c^2 + 2ab$
$a^2 + 2ab + b^2 = c^2 + 2ab$
$a^2 + b^2 = c^2$

In the twelfth century, Bhaskara, a Hindu mathematician, made this proof:

Place the right triangle and three copies of the given right triangle in the following arrangement:

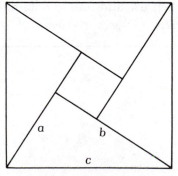

The area of the large square is $c^2$
The area of one right triangle is $\frac{1}{2}ab$
The area of the small square is $(b - a)^2$ (Do you see this?)
So the large square = 4 triangles + the small square
$c^2 = 4(\frac{1}{2}ab) + (b - a)^2$
Which simplifies to $c^2 = a^2 + b^2$

It should be mentioned that Euclid did much more than just copy the theorem that Pythagoras had proven several hundred years earlier. He extended it to prove that if similar figures *of any kind* are drawn on the three sides of a right triangle, the area of the figure on the hypotenuse is equal to the sum of the areas on the two legs!

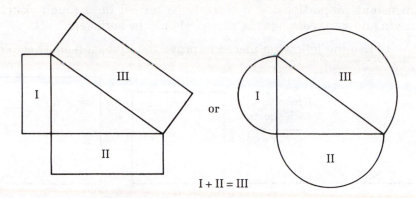

I + II = III

Or even

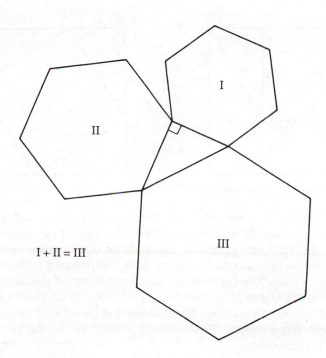

I + II = III

This set of exercises is more activity oriented than most in this book. Most of these problems will make excellent puzzles, if the pieces are cut from a stiff material such as cardboard, wood, or plastic. If the figure is placed on top of the heavier material and pinholes are made at the vertices, this will leave a pattern for cutting the heavier material. If you wish to construct a larger copy of one of these figures, be sure to make the dimensions proportional. For work to be turned in, a rough sketch, showing how to rearrange the pieces, should be sufficient.

1. Use the following pieces to prove the Pythagorean Theorem. $(a^2 + b^2 = c^2)$

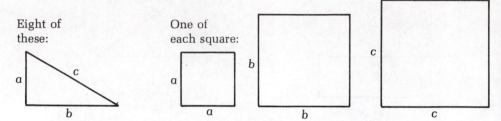

Eight of these:

One of each square:

*Hint:* Arrange the pieces to form these two equal squares, and then subtract the four triangles from each.

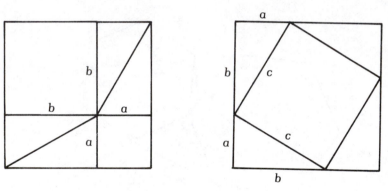

2. This problem not only demonstrates the Pythagorean Theorem, but gives a method of rearranging the material of two squares into a single square. As the story is told, this method was devised by a Russian furrier. It seems the Czar had a square hole, $c$ units on a side, in his rare blue ermine cloak. The furrier had just enough material to fill the hole, but it was in the shape of two squares, $a$ units and $b$ units on a side. The problem was to cut them in such a manner that they could be rearranged to form one large square. The method is this: place the two squares adjacent to each other.

Mark off $b$ across the bottom of $a$ and $b$.

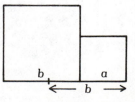

Draw the diagonals shown

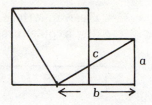

Now rearrange the five pieces into a square with a side of $c$, showing that $a^2 + b^2 = c^2$.

3. Leonardo da Vinci has been credited with this approach: begin with the usual figure

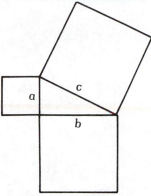

Form two additional triangles congruent to the original one by drawing $VWX$ and $QR$

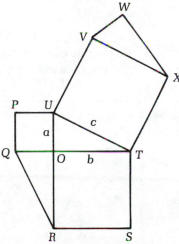

Now draw $PS$, and reflect $PSTU$ to this configuration by putting $P$ at $S$ and $S$ at $P$.

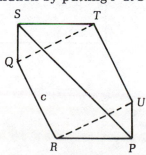

Show that this is congruent to $OUVWXT$. Now, by subtracting two triangles from each area, show $a^2 + b^2 = c^2$.

4. Rearrange the parts of the squares on the legs to form a square on the hypotenuse.

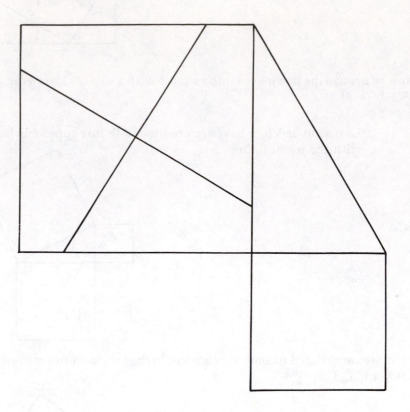

5. Rearrange the parts of the squares on the legs to form a square on the hypotenuse.

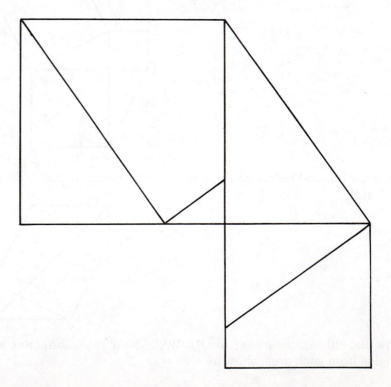

6. Rearrange the parts of the squares on the legs to form a square on the hypotenuse.

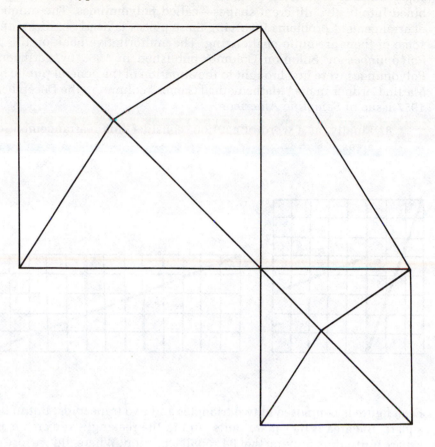

7. Rearrange the parts of the squares on the legs to form a square on the hypotenuse.

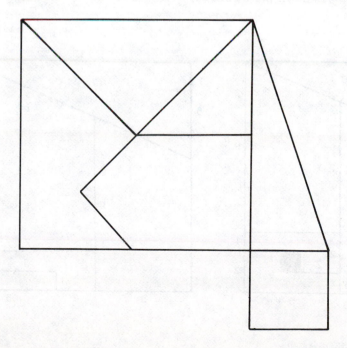

If you enjoy these rearrangement type puzzles, then you would probably enjoy working with polyominoes. Five one-unit squares can be combined into twelve different shapes—called **polyominoes.** The number of arrangement problems involving these pieces is nearly endless, and some of them are quite challenging. The authoritative book on this is *Polyominoes* by Solomon Golomb, published in 1965 by Scribners. Polyominoes were first brought to the attention of the general public by Martin Gardner in his "Mathematical Games" column in the December, 1957 issue of *Scientific American.*

8. Finally, as a word of caution, consider this rearrangement of areas:

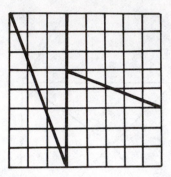

Each figure is composed of two triangles and two trapezoids. But in the square, the area is 64 square units, and in the rectangle, the area is 65 square units; thus proving that 64 = 65!! Question: Where did we lose or gain one unit?

9. Here is one a little more difficult. It demonstrates that 143 = 144! Explain the difficulty.

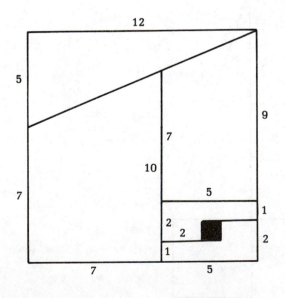

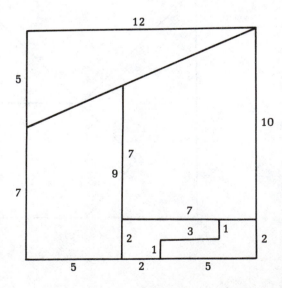

10. If angle $C$ is a right angle, then by the Pythagorean Theorem, $a^2 + b^2 = c^2$.

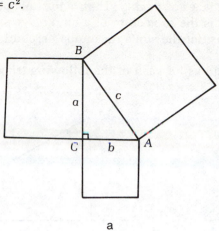

a

What will happen to this relationship if $a$ and $b$ remain the same length, and angle $C$ is decreased to an acute angle?

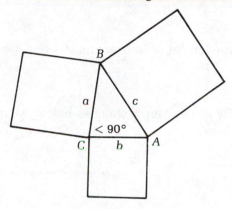

b

$$a^2 + b^2 \; ? \; c^2 \qquad (<, >, \text{ or } =)$$

What happens if $a$ and $b$ remain the same length, and angle $C$ is increased to an obtuse angle?

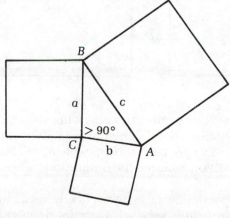

c

$$a^2 + b^2 \; ? \; c^2$$

In each of these cases, we need some "adjustment term" on the left side of the equation to make an equality. That "adjustment term" turns out to be $2a'b$, where $a'$ is the projection of $a$ onto $b$. (A projection is like a shadow, if we consider the sun's rays to be perpendicular to $b$.)

Complete the formulas for each of the following triangles:

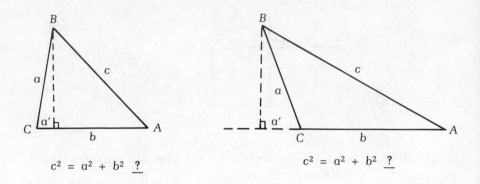

$$c^2 = a^2 + b^2 \underline{\ ?\ }$$

$$c^2 = a^2 + b^2 \underline{\ ?\ }$$

These formulas are developed in trigonometry and are called the Law of Cosines.

## 30. The golden mean

Look carefully at the four rectangles below. Which has the most pleasing proportions?

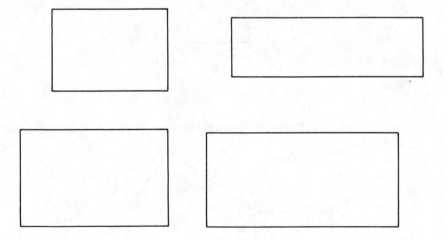

Most people will chose the third one. It has sides in the ratio of about 5 to 8. This ratio is called the golden ratio, the golden mean, or the golden rectangle. Seventy-five per cent of the people tested by Wilhelm Wundt and others (circa 1876) preferred the proportions of the golden rectangle to others.

Why is this particular ratio aesthetically pleasing? One theory is that it is a pleasant ratio because it occurs frequently in nature and people are comfortable with it. More about its natural occurrences later.

This ratio, which the Greeks "discovered", is derived in this manner:

We wish to divide a rectangle into two parts, in such a manner that we have a square and a new rectangle similar to the original one. That is, the ratio of the length to width in the original rectangle is the same as the ratio of the length to width in the rectangle remaining after a square is removed.

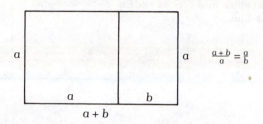

So doing some algebra:

$$a^2 = ab + b^2 \qquad \text{Prod. of means = prod. of extremes}$$

$$a^2 - ab - b^2 = 0$$

$$a = \frac{b \pm \sqrt{b^2 - 4(1)(-b^2)}}{2} \qquad \text{Quadratic formula}$$

$$a = \frac{b \pm \sqrt{5b^2}}{2}$$

$$a = \frac{b \pm b\sqrt{5}}{2}$$

$$a = \frac{1 \pm \sqrt{5}}{2}b$$

Since we are not interested in negative distances

$$a = \frac{1 + \sqrt{5}}{2}b \qquad \text{or} \qquad a = \phi b \qquad \text{or} \qquad \frac{a}{b} = \phi$$

Since we will be using this ratio often, we will give it the name $\phi$ (the Greek letter phi). This is the golden ratio. If we calculate a decimal approximation $\phi \approx 1.618033989 \ldots$ Or, if we solve the equation above for $b$, $b = \frac{a}{\phi}$ or, $\frac{b}{a} = \frac{1}{\phi}$. What is this decimal approximation?

$\frac{1}{\phi} \approx .618033989 \ldots$ Notice the interesting relationship between $\phi$ and $\frac{1}{\phi}$! One is the decimal part of the other. If you think that relationship is curious, try calculating $\phi^2$.

Do you remember the construction of the regular pentagon? Part of that construction was the construction of a mean proportional. Notice in the derivation above that $a$ is the mean proportional to $a + b$ and $b$. However, our problem here is a little different than that of the pentagon construction. Instead of having $AB$ in the proportion $\frac{AB}{AC} = \frac{AC}{BC}$ and finding $AC$, we have $AC$ and wish to find $AB$.

The construction can be done this way:

**Mark off $AC$ on a working line**
**Construct a perpendicular to $AC$ at $C$**
**Mark off $CD = AC$**
**Bisect $AC$, call the midpoint $E$**
**Now with $E$ as center and $DE$ as radius**
**Locate $B$ on the line**

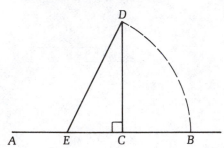

Let us see how this works out numerically, if we let $AC = 1$

$AC = 1, CD = 1, CE = \frac{1}{2}, AE = \frac{1}{2}$
$DE^2 = CE^2 + CD^2 = (\frac{1}{2})^2 + 1^2 = \frac{5}{4}$        by Pythagorean Theorem
so $DE = \dfrac{\sqrt{5}}{2}$
$AB = AE + EB$
and $EB = DE$
so $AB = AE + DE = \frac{1}{2} + \dfrac{\sqrt{5}}{2}$
$= \dfrac{1 + \sqrt{5}}{2} = \phi\,!!$

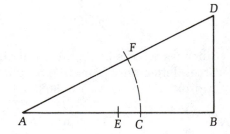

While we are at it, let's look at that construction of the mean pro-portional numerically. If you recall the construction, we are given $AB$ and wish to find $AC$ such that $\dfrac{AB}{AC} = \dfrac{AC}{BC}$ .

**Bisect $AB$, call it $E$**
**Construct a perpendicular to $AB$ at $B$**
**Mark off $BD = BE$**
**Draw $AD$**
**Mark off $DF = BD$**
**With $A$ as center and $AF$ as radius**
**Locate $C$ on $AB$**

Now if we let $AB = 1$, $AE = BE = BD = DF = \frac{1}{2}$

$$AD^2 = AB^2 + BD^2 = 1^2 + (\tfrac{1}{2})^2 = \tfrac{5}{4}$$

$$AD = \frac{\sqrt{5}}{2}$$

$$AC = AF = AD - DF = \frac{\sqrt{5}}{2} - \frac{1}{2} = \frac{\sqrt{5} - 1}{2} = \frac{1}{\phi}\,!!$$

(To see that this is equal to $\frac{1}{\phi}$ calculate a decimal approximation, or better yet, calculate $\frac{1}{\phi}$ by rationalizing the denominator.)

Back to our golden rectangle. To construct a rectangle with the best proportions, when a side (or square) is given:

**Bisect a side.**
**From the point of bisection, using the distance to a non-adjacent vertex as radius, swing an arc to determine the length of the rectangle.**

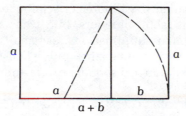

Now since $\frac{a+b}{a} = \frac{a}{b}$, the small rectangle is similar to the original. If we again cut off a square, then the new rectangle formed is similar to the original, and we could continue this process indefinitely!!

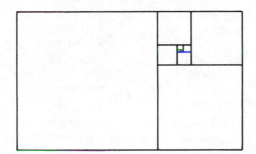

Now an interesting thing happens, if we connect the corners of the successive squares formed with a smooth curve. We get a nice spiral. It is called a logarithmic spiral, and is very common in nature.

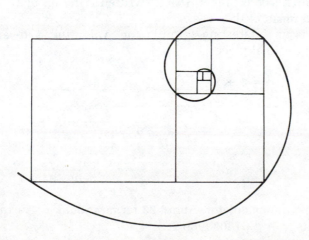

Now let us discuss this number $\phi$ in a different manner. The greatest medieval mathematician, Leonardo of Pisa, nicknamed Fibonacci, introduced the sequence of numbers, 1, 1, 2, 3, 5, 8, 13, 21, 34, 55, 89, 144, 233, . . . way back in the year 1202 AD. It is said that this sequence came about as a response to a problem about the multiplication of rabbits. We begin with one pair of rabbits on January 1, they cannot reproduce in their first month of life, but produce one new pair each month thereafter. Assuming no deaths, the number of pairs of rabbits we have each month thereafter is 1, 1, 2, 3, 5, 8, 13, 21, 34, 55, 89, and at the end of one year, 144. Can you draw some kind of diagram to show how this would work?

Algebraically, we see that each term of the sequence is the sum of the two terms that precede it. $A_n = A_{n-1} + A_{n-2}$

This sequence, called the Fibonacci sequence, may seem of little interest to anyone but rabbit breeders, but it has so many interesting ramifications and so many applications, especially in nature, that for years now there has been a magazine published every three months devoted strictly to the Fibonacci sequence. It is appropriately called the *Fibonacci Quarterly*.

Consider the *ratio* of successive terms of the Fibonacci sequence:

| Ratio | Value |
|-------|-------|
| 2/1 | 2 |
| 3/2 | 1.5 |
| 5/3 | 1.666666 |
| 8/5 | 1.600000 |
| 13/8 | 1.625000 |
| 21/13 | 1.615385 |
| 34/21 | 1.619048 |
| 55/34 | 1.617647 |
| 89/55 | 1.618182 |
| 144/89 | 1.617978 |
| 233/144 | 1.618056 |
| 377/233 | 1.618026 |
| 610/377 | 1.618037 |

What value do you think this ratio is approaching? The limit, as $n$ gets bigger and bigger of $\dfrac{A_n}{A_{n-1}} = \phi$!! So, this sequence introduced by Fibonacci in his book "Liber Abaci," written at the age of 27, ties in with the golden mean of the Greeks.

Here is an interesting formula for evaluating $\phi$ to any desired number of decimal places:

$$\phi = 1 + \cfrac{1}{1 + \cfrac{1}{1 + \cfrac{1}{1 + \cfrac{1}{1 + \cfrac{1}{1 + \cfrac{1}{\text{and so on}}}}}}}$$

Try this on your calculator. About 23 terms should give a value correct to 10 digits, $\phi \approx 1.618033989 \ldots$

Here is another one that converges a little more rapidly:

$$\phi = 1 + \sqrt{1 + \cfrac{}{\sqrt{1 + \cfrac{}{\sqrt{1 + \cfrac{}{\sqrt{1 + \cfrac{}{\sqrt{1 + \text{ and so on}}}}}}}}}}$$

Twenty terms on this one should give ten digit accuracy. One has to wonder why the same basic pattern, with either reciprocals or radicals, gives this number $\phi$!

We have mentioned that there are many applications of the number $\phi$, the golden rectangle, the spiral formed from the golden rectangle, and the Fibonacci sequence—so let us now list some examples.

*"The Parthenon," Nashville.*

The Parthenon, built in Athens in about the 5th century BC, has been considered one of the world's most beautiful buildings. The ratio of width to height of its front, the height of columns to their distance apart, and many other parts, is the golden mean. The ratio occurs in the dimensions of the pyramids. The ratio of their width to height is very nearly the golden mean. The ratio $\phi$ is often used even today by modern architects in the construction of buildings. The ratio can be seen in the constructions of the architect Le Corbusier, who suggested that our lives are "comforted" by mathematical shapes.

**ARCHITECTURE**

Well-proportioned vases have for
many years used the golden mean for
the ratio of their height to width.
Shown here: Cypriote amphora (c.600
BC), Chalcidian amphora, Greek vase
(515 BC), Apache storage basket
(c.1950 AD), jar from Ch'ing Dynasty
(1662–1722 AD), jar from Japan Asuka
Period (552–645 AD). *Courtesy Leland
Stanford, Jr. Museum.*

The ratio of height to width of a well-proportioned vase is $\phi$; a great many of the Greek vases had this ratio for dimensions. Many famous painters have used the golden mean. Leonardo da Vinci not only used it, but taught it and wrote about it. His painting "Saint Jerome" fits almost perfectly into a golden rectangle. Considering his love of geometry and mathematics in general, it would be expected that he would use this principle. The ratio can also be found as a part of Michelangelo's "Holy Family," Botticelli's "Magnificat," Salvador Dali's "Corpus Hypercubus," "La Parade" by the French Impressionist Georges Seurat, and much of Piet Mondrian's work, especially "Place de la Concorde"—a linear abstraction. Juan Gris not only used the golden mean, but specifically praised its virtues.

Chambered nautilus.

Starfish, sand dollars, and sea anemones all use the pentagon and the ratio of the golden mean.

The epeira spider follows a logarithmic curve in building its web.

The logarithmic spiral occurs in the Chambered Nautilus, snails, elephant tusks, the horns of wild sheep, a lion's claws, a parrot's beak, and even the claws of a canary. The spiral can also be seen in the web of the epeira spider (a common garden spider). The ratio $\phi$ can be found in several places in the starfish (see discussion of pentagram). The Fibonacci sequence is found in the lineage of male bees, and in bacterial growth, as well as the rabbits we've already discussed.

The number of "leaves" each of the spirals passes through on a pine cone is usually a number in the Fibonacci sequence.

Similarly to the preceding figures, the spirals of a pineapple generally pass through a number of leaves that is from the Fibonacci sequence.

**BOTANY**

The number of seeds or leaves in the whorls of pine cones, sunflowers, and pineapples are numbers of the Fibonacci sequence. In phyllotaxis—the arrangement of leaves along a stem—the number of rotations and the number of buds until one is directly above another are numbers of the Fibonacci sequence. Many varieties of trees have $\phi$ as the ratio of their height to width. This seems to promote the most efficient gathering of light and moisture.

Likewise, the curves visible in the seed head of a sunflower pass through a number of seeds that is determined by the Fibonacci sequence.

The applications are really too numerous to mention—but, we have already indicated a few with our discussions of the mean proportion, construction of pentagon, sequences, and spirals. As you learn more mathematics, you seem to find more and more places where $\phi$ pops up. As an additional example, consider the pentagram—the symbol of the Society of Pythagoreans (also applicable to the starfish mentioned on page 373):

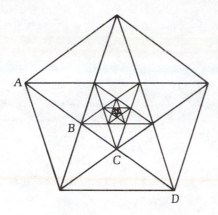

On each side of each star, we have the ratio $\phi$ in several places.

$$\frac{AB}{BC} = \phi \text{ and } \frac{AC}{AB} = \phi \text{ and } \frac{AD}{AC} = \phi$$

Additionally, if the points of the star are folded up to form a regular pentagonal pyramid, the ratio of the pyramid's height to the radius of its base, $\frac{AO}{OB} = \phi$

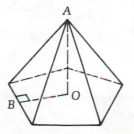

All these ratios of $\phi$ might be somewhat expected if we recall that part of the construction of the regular decagon and pentagon involved the construction of a mean proportional.

The entire musical scale is based on ratios of tones. The Pythagoreans were very much interested in the relationship of these various ratios. The chord which has sometimes been called the most satisfying musical chord (like the most pleasing rectangle) is the major sixth *E*—which has a ratio of 5 to 8 to *C*.

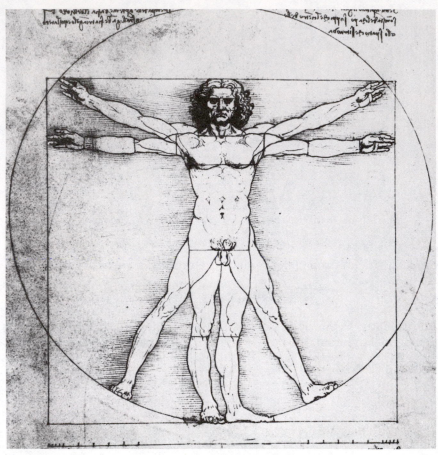

There are many ways to measure DaVinci's "Man" that produce the ratio of the golden mean.

**PEOPLE (ANATOMY)**

In 1509, Leonardo da Vinci wrote a book with a monk, Luca Pacioli, called *De Divina Proportione* (The Divine Proportion). In its discussion of figure sculpture, they proposed that a perfectly proportioned body should have the golden mean in several ratios. All of the following measurements should have a ratio of $\phi$:

Total height : height of navel from floor
Navel to top of head : armpit to top of head
Widest part of thigh : narrow part of thigh
Width of head : width of throat
Width of calf : width of ankle
Width of forearm : width of wrist
Height of head : width of head
Chin to eyes : chin to nose
Eyes to mouth : eyes to nose

A breaking wave typically has the shape of a logarithmic spiral. *"Over the Falls" by Gary Patterson.*

Of course, if you are thin, heavy, long-waisted, have a round face, or do not have a navel, then these proportions will not apply to you as an individual. Perhaps beauty contests should be judged by Leonardo's proportions!

The shape of a spiral nebula is the logarithmic spiral.

Surfers shooting the gap underneath a breaking wave are covered by a logarithmic spiral of ocean spray. The shore line of Cape Cod is a portion of a spiral. And as long as we are on a nautical theme, the best shape for the arms of an anchor is the same spiral. The logarithmic spiral can be found in the spiral shape of a galaxy and in the shape of a comet's tail. And from the sublime to the ridiculous—the shape of some playing cards is a golden rectangle.*

Finally, a quote from Johann Kepler:

> Geometry has two great treasures: one is the theorem of Pythagoras; the other is the division of a line into extreme and mean ratio (the golden mean). The first, we may compare to a measure of gold; the second, we may name a precious jewel.

**EXERCISE SET 30**

1. Move a ruler along in a vertical position on this figure until the rectangle formed is the shape that you like best. Draw a line to complete the rectangle. Now measure its length and width, and divide to find the ratio.

2. Construct a square with a 10 cm side. Now use that square to construct a golden rectangle.

3. An artist wishes to frame a canvas in the shape of a golden rectangle. If the canvas has a width of 36 cm, what should its length be? (Give your answer to the nearest tenth of a centimeter.)

4. If the canvas above is 64 cm long, how wide should it be made?

5. A normal sheet of paper is $8\frac{1}{2}$ in. by 11 in. Is its length too long or too short for the golden rectangle?

6. A sheet of legal-sized paper is $8\frac{1}{2}$ in. by 14 in. Is it too long or too short for the golden rectangle?

7. A Lucas sequence is very much like a Fibonacci sequence, except that it can begin with *any* two numbers. Each successive term is then the sum of the preceding two terms. Choose any two numbers, and then form eight more terms of a Lucas sequence. Find the ratio of the 6*th* term divided by the 5*th*, the 7*th* divided by the 6*th*, and so on to the 10*th* divided by the 9*th*. What conclusion do you make?

8. A student wishes to make a model of the Parthenon. It is to have a 54 cm base. What will be its height?

9. A publisher wishes to have a geometry book in the most pleasing shape. If the width of the book is 17 cm, what will be its height?

10. A vase is to be made 30 cm high. In order to have the height to width ratio equal to $\phi$, what should be its width? (to the nearest cm)

11. In the pentagon on page 375 if $AD = 10$ cm, find $AC$, $AB$, and $BC$.

---

*Much additional information can be found in *The Divine Proportion*, by H. E. Huntley—a 1970 Dover publication, and in *The Geometry and Art of Life*, by Matila Ghyka—also from Dover (1977).

12. An artist is drawing a 22 cm high figure of a person. What should be the height of the navel from the floor? How far from the navel to the top of the head? How far from the armpit to the top of the head?

13. An artist is drawing a portrait in which the head is 12 cm high. What should be the width of the head?

14. An artist is drawing a portrait in which the distance from the chin to the eyes is 11 cm. Find the proper distance from chin to nose, from eyes to nose, and from eyes to mouth.

15. A carpenter wishes to construct some cupboards that have $\phi$ as their height to width ratio. If they need to be 60 cm wide, how high should they be?

16. An architect wishes to design some entry doors to a patio with the ratio of $\phi$ as their height to width ratio. The doors need to be 200 cm tall. How wide should they be?

17. Choose five of the nine ratios of body proportions given by Leonardo da Vinci and calculate these ratios for yourself and a friend. (It may be difficult to measure yourself.) For the measurements of width of thigh, calf, forearm, and wrist it is easier to use circumference rather than width—both numerator and denominator of the ratio will contain a factor of $\pi$, so the ratio will be the same. Measure as carefully as you can, but don't be surprised if some ratios are not too close to $\phi$.

18. Astronomers have found that some solar and lunar eclipses show certain repeating behavior patterns every 6, 41, 47, 88, . . . years. What kind of a sequence is this? Write the first ten terms of this sequence.

19. Add the first three Fibonacci numbers. Add the first four, the first five, the first six, and so on. Do you see a pattern? Write a formula for $S_n$ (the sum of the first $n$ terms), as a function of one of the terms of the sequence.

---

Different parallel postulates:

8—Summary

Euclid—Through a point not on a line, there is exactly one line parallel to the given line.
Lobachevski—Through a point not on a line, there is more than one line parallel to the given line.
Riemann—Through a point not on a line, there are no lines parallel to the given line.

The sum of the angles of a triangle:

Euclid—exactly 180°
Lobachevski—less than 180°
Riemann—more than 180°

Concurrency theorems:

The angle bisectors of a triangle are concurrent. (incenter)

The perpendicular bisectors of the sides of a triangle are concurrent. (circumcenter)

The altitudes of a triangle are concurrent. (orthocenter)

The medians of a triangle are concurrent in a point $\frac{2}{3}$ of the distance from any vertex to the opposite side. (centroid, or center of gravity)

The Pythagorean Theorem as an area relationship:

In right-angled triangles, the square on the side subtending the right angle is equal to the squares on the sides containing the right angle.

The golden mean:

$$\phi = \frac{1 + \sqrt{5}}{2} \approx 1.618$$

The Fibonacci sequence:

1, 1, 2, 3, 5, 8, 13, 21, 34, 55, 89, 144, 233, . . .

where $A_n = A_{n-1} + A_{n-2}$

---

## Epilogue

*Pure mathematics consists entirely of such assertions as that, if such and such a proposition is true of anything, then such and such another proposition is true of that thing. It is essential not to discuss whether the first proposition is really true, and not to mention what the anything is of which it is supposed to be true . . . If our hypothesis is about anything, and not about some one or more particular things, then our deductions constitute mathematics. Thus, mathematics may be defined as the subject in which we never know what we are talking about, nor whether what we are saying is true.*

—Bertrand Russell

# APPENDIX I

# APPENDIX II

| POSTULATES | Page |
|---|---|
| 1. One and only one straight line can be drawn through two given points. (Two points determine a line.) | 38 |
| 2. Only one plane can be passed through three non-collinear points. (Three points determine a plane.) | 38 |
| 3. To every pair of points there corresponds a unique positive number. (Distance Postulate) | 39 |
| 4. To every angle there corresponds a unique positive number. (Angle Measure Postulate) | 40 |
| 5. There is a one-to-one correspondence between a set of points and their reflection about a line. (Point Reflection Postulate) | 42 |
| 6. Through a point not on a line only one parallel to the line can be drawn. (Euclidean Parallel Postulate) | 43 |
| 7. Any property of regular polygons which does not depend on the number of sides of the polygon is also true of circles. (Limit Postulate) | 43 |
| 8. The existence of the transformations. | 54 |

# APPENDIX III

| THEOREMS AND COROLLARIES | Page |
|---|---|
| If two straight lines intersect, the vertical angles are equal. | 64 |
| If two straight lines intersect to form equal adjacent angles, then the lines are perpendicular. | 65 |
| If two triangles have two sides and the included angle of one equal respectively to two sides and the included angle of the other, then they are congruent. SAS | 76 |
| If two right triangles have their legs respectively equal, then the triangles are congruent. | 78 |
| The bisector of the vertex angle of an isosceles triangle is the perpendicular bisector of the base. | 80 |
| If two triangles have two angles and the included side of one equal respectively to two angles and the included side of the other, then they are congruent. ASA | 81 |
| If two right triangles have a leg and adjacent angle respectively equal, then they are congruent. | 82 |
| If two sides of a triangle are equal, then the angles opposite those sides are equal. | 87 |
| If two angles of a triangle are equal, then the sides opposite those angles are equal. | 88 |
| An equilateral triangle is also equiangular. | 89 |
| An equiangular triangle is also equilateral. | 90 |
| If two triangles have three sides of one equal to three sides of the other, then they are congruent. SSS | 91 |
| Two lines in the same plane perpendicular to the same line are parallel. | 112 |
| Two lines in the same plane parallel to the same line are parallel to each other. | 112 |
| If two lines are cut by a transversal so that the alternate interior angles are equal, then the lines are parallel. | 114 |
| If two parallel lines are cut by a transversal, then the alternate interior angles will be equal. | 116 |
| If two parallel lines are cut by a transversal, then the corresponding angles will be equal. | 116 |
| The sum of the angles of a triangle is 180°. | 126 |

*Theorems and corollaries continue on the following page*

| | |
|---|---|
| The base angles of an isosceles trapezoid are equal. | 199 |
| The diagonals of an isosceles trapezoid are equal. | 200 |
| The opposite sides of a parallelogram are equal. | 201 |
| The opposite angles of a parallelogram are equal. | 202 |
| The diagonals of a parallelogram bisect each other. | 202 |
| Parallel lines are everywhere equidistant. | 203 |
| If the opposite sides of a quadrilateral are equal, it is a parallelogram. | 204 |
| If the opposite angles of a quadrilateral are equal, it is a parallelogram. | 205 |
| If the diagonals of a quadrilateral bisect each other, it is a parallelogram. | 206 |
| If a quadrilateral has one pair of sides both equal and parallel, then it is a parallelogram. | 207 |
| If the diagonals of a parallelogram are equal, then it is a rectangle. | 208 |
| If the diagonals of a quadrilateral are perpendicular bisectors of each other, then it is a rhombus. | 209 |
| If the diagonals of a quadrilateral are equal and perpendicular bisectors of each other, then it is a square. | 210 |
| The area of a square is a side squared. | 220 |
| The area of a parallelogram is its base times its height. | 221 |
| The area of a triangle is $\frac{1}{2}$ its base times its height. | 222 |
| Triangles with equal bases and equal altitudes are equal. | 223 |
| The median of a triangle divides it into two equal triangles. | 224 |
| In a right triangle, the product of the legs equals the product of the hypotenuse and the altitude to the hypotenuse. | 225 |
| The area of an equilateral triangle with side $s$ is $\dfrac{s^2 \sqrt{3}}{4}$. | 226 |
| The area of a triangle is one-half its perimeter times the radius of its inscribed circle. $A = \frac{1}{2}pr$ | 227 |
| The area of a rhombus is one-half the product of the diagonals. $A = \frac{1}{2}d_1 d_2$ | 228 |
| The area of any triangle is $\sqrt{s(s-a)(s-b)(s-c)}$ where $s$ is the semiperimeter. (Heron's semiperimeter formula) | 228 |
| The area of a trapezoid is one-half the product of its height and the sum of its bases. $A = \frac{1}{2}h(b_1 + b_2)$ | 229 |
| The area of a regular polygon is one-half its apothem times its perimeter. $A = \frac{1}{2}ap$ | 231 |
| Any three noncollinear points determine a circle. | 246 |

*Theorems and corollaries continued on the following page*

In the same, or in equal, circles equal central angles have equal arcs. 247

In the same, or in equal, circles equal arcs have equal central angles. 248

In the same, or in equal, circles equal arcs have equal chords. 249

In the same, or in equal, circles equal chords have equal arcs. 250

In the same, or in equal, circles equal central angles have equal chords. 250

In the same, or in equal, circles equal chords have equal central angles. 251

A line through the center of a circle, perpendicular to a chord, bisects the chord and its arcs. 252

The perpendicular bisector of a chord passes through the center of the circle. 253

In the same or in equal circles, equal chords are equidistant from the center. 254

In the same or in equal circles, chords equidistant from the center are equal. 255

The perpendicular to a radius at its extremity is a tangent. 256

A radius drawn to the point of contact of a tangent is perpendicular to the tangent. 257

Tangents to a circle from an external point are equal. 258

Two parallel lines intercept equal arcs on a circle. 261

An inscribed angle is half the measure of its intercepted arc. 270

Angles inscribed in the same or in equal arcs are equal. 270

An angle inscribed in a semicircle is a right angle. 271

An angle formed by two intersecting chords is half the sum of the two intercepted arcs. 275

An angle formed by a chord and a tangent at one end of the chord is half its intercepted arc. 276

An angle formed by two secants, a secant and a tangent, or by two tangents is half the difference of the intercepted arcs. 277

If a tangent and a secant are drawn to a circle from an external point, the tangent is the mean proportional to the secant and its external segment. 280

The area of a circle is pi times the radius squared. 288

$$A = \pi r^2$$

The area of a sector of a circle is the measure of its central angle divided by 360, times the area of the circle. 290

$$A = \frac{C}{360} \pi r^2$$

The area of a segment of a circle is the area of its sector minus its triangle.                                291

The area of a ring is the difference of the areas of its circles.                                292

The angle bisectors of a triangle are concurrent. (incenter)                                338

The perpendicular bisectors of the sides of a triangle are concurrent. (circumcenter)                                338

The altitudes of a triangle are concurrent. (orthocenter)                                339

The medians of a triangle are concurrent in a point $\frac{2}{3}$ of the distance from any vertex to the opposite side. (centroid, or center of gravity)                                340

In right-angled triangles, the square on the side subtending the right angle is equal to the squares on the sides containing the right angle. (Pythagorean Theorem)                                355

## SURFACE FORMULAS

| | |
|---|---|
| Rectangular parallelepiped | $2(lh) + 2(wh) + 2(lw)$ |
| Cube | $6a^2$ |
| Prism | $ph + 2B$ |
| Circular cylinder (right or oblique) | $2\pi r(h + r)$ |
| Pyramid (right regular) | $\frac{1}{2}pl + B$ |
| Right circular cone | $\pi r(l + r)$ |
| Frustum of a cone | $2\pi r_M l$ |
| Sphere | $4\pi r^2$ |

## VOLUME FORMULAS

| | |
|---|---|
| Rectangular parallelepiped | $lwh$ |
| Cube | $a^3$ |
| Any parallelepiped | $Bh$ |
| Any prism | $Bh$ |
| Any cylinder | $Bh$ |
| Right circular cylinder | $\pi r^2 h$ |
| Any pyramid | $\frac{1}{3}Bh$ |
| Any cone | $\frac{1}{3}Bh$ |
| Right circular cone | $\frac{1}{3}\pi r^2 h$ |
| Sphere | $\frac{4}{3}\pi r^3$ |

# APPENDIX IV

## GLOSSARY

**acute angle**   an angle with degree measure less than 90°.

**adjacent angles**   angles having the same vertex and a common ray between them.

**alternate exterior angles**   angles on alternate sides of a transversal that are also exterior to the two lines the transversal intersects.

**alternate interior angles**   angles on alternate sides of a transversal that are also between the two lines the transversal intersects.

**angle**   figure formed by two rays with a common endpoint.

**apothem**   the line from the center of a regular polygon that is perpendicular to a side.

**arc**   a portion of the circumference of a circle.

**auxiliary line**   a line added as an aid in a proof.

**axiom**   a statement accepted without proof.

**bisect**   to divide into two equal parts.

**central angle**   an angle with its vertex at the center of a circle.

**centroid**   the center of gravity of a triangle, its "balance point," the point of concurrency of the medians.

**chord**   a line segment connecting two points of a circle.

**circle**   a closed curve in a plane, all points of which are equidistant from a point called the center.

**circumcenter**   a point which is the center of a circle circumscribed about a triangle.

**circumference**   the length of the curve of a circle.

**circumscribed**   a figure drawn around another figure.

**coincidence**   a method of indirect proof in which two figures are shown to coincide.

**collinear**   lying on the same line.

**common external tangent**   a line tangent to two circles that does not cut through their line of centers.

**common internal tangent**   a line tangent to two circles that cuts through their line of centers.

**common tangent**   a line segment tangent to two or more circles.

**complementary angles**   two angles which can form a right angle; two angles whose degree measure adds to 90°.

**concave polygon**   a polygon in which at least one of the interior angles has measure greater than 180°.

**concentric**   two or more circles which have the same center.

**conclusion**   the "prove" or "then" part of a statement.

**concurrent**   two or more lines which pass through the same point.

**congruent**   figures of the same size and shape.

**contrapositive**   a statement formed by taking the inverse of the converse of a given statement.

**converse**   a statement formed by interchanging the hypothesis and conclusion of a given statement.

**convex polygon**   a polygon in which all of the interior angles are less than 180°.

**corresponding angles**   pairs of angles having the same relative positions in two figures.

**decagon**   a ten-sided polygon.

**deductive logic**   reasoning proceeding from a generally accepted statement to a specific case.

**diagonal**   line joining two non-adjacent vertices.

**diameter**   a line segment through the center of a circle terminated at both ends by the circle.

**dilatation**   a figure formed by a transformation in which the corresponding sides of figures are parallel.

**disjoint sets**   sets which have no elements in common.

**dodecagon**  a twelve-sided polygon.

**dodecahedron**  a polyhedron with twelve faces.

**equiangular**  having all angles equal.

**equilateral**  having all sides equal.

**Euler circles**  circles used to represent sets, also called Venn diagrams.

**exclusion**  a method of indirect proof where all but one of several possibilities are eliminated.

**exterior angle of a triangle**  an angle formed by one of the sides of the triangle and an adjacent side extended.

**frustum of a cone**  portion of a cone cut off by two parallel planes.

**heptagon**  a seven-sided polygon.

**hexagon**  a six-sided polygon.

**hexahedron**  a polyhedron with six faces.

**hypotenuse**  the side opposite the right angle in a right triangle.

**hypothesis**  the "given" or "if" part of a statement.

**icosahedron**  a polyhedron with twenty faces.

**incenter**  the point in a triangle which is the center of the inscribed circle.

**inductive logic**  reasoning from the specific to the general.

**inscribed**  a figure drawn inside another figure.

**inscribed angle**  an angle with its vertex on the circumference of a circle, and with chords or secants for its rays.

**inverse**  a statement formed by negating both the hypothesis and conclusion of a given statement.

**isosceles**  having two sides of the same length.

**isosceles trapezoid**  a trapezoid with its non-parallel sides equal.

**kite**  a quadrilateral with two pairs of equal consecutive sides, each pair of a different length.

**line of centers**  a line connecting the centers of two circles.

**line segment**  the portion of a line between two points.

**major arc**  an arc of a circle that is greater than a semicircle.

**mean proportional**  a quantity used as both the second and third terms in a proportion; the $X$ in $\dfrac{a}{X} = \dfrac{X}{b}$

**median**  a line in a triangle from a vertex to the mid-point of the opposite side.

**minor arc**  an arc of a circle that is less than a semicircle.

**nonagon**  a nine-sided polygon.

**noncollinear**  not lying on the same line.

**oblique prism**  a prism in which the lateral edges are not perpendicular to the plane of the base.

**obtuse angle**  an angle with degree measure more than 90° but less than 180°.

**octagon**  an eight-sided polygon.

**octahedron**  a polyhedron with eight faces.

**orthocenter**  the point of concurrency of the altitudes of a triangle.

**parallel**  lines lying in the same plane that do not intersect.

**parallelogram**  a quadrilateral with its opposite sides parallel.

**pentagon**  a five-sided polygon

**perimeter**  the sum of the lengths of the sides of a polygon.

**perpendicular**  lines which form a right angle, or 90°.

**phi ($\phi$)**  the ratio of the golden mean.

**pi ($\pi$)**  the ratio of the circumference of a circle to its diameter.

**polygon**  a closed plane figure composed of three or more line segments.

**polyhedron**  a solid which has polygons for each of its faces.

**postulate**  a statement accepted without proof.

**prism**  a solid figure formed when the corresponding vertices of two congruent figures that are lying in parallel planes are joined.

**proof**  some demonstration that convinces the observer.

**proportion**  the equality of two ratios.

**pyramid**  a solid figure formed by joining each vertex of a polygon to some point not in the plane of the polygon.

**quadrilateral**  a polygon with four sides.

**radius**  the distance from the center of a circle to a point on its circumference.

**ratio**  the comparison of two numbers by division.

**rectangle**  a parallelogram with a right angle.

*reductio ad absurdum*  a method of indirect proof in which one of two possibilities is eliminated.

**reflection**   $P'$ is a reflection of $P$ through line $L$, if and only if $L$ is the perpendicular bisector of $PP'$.

**reflex angle**   an angle with degree measure greater than 180°.

**regular polygon**   a polygon which is both equilateral and equiangular.

**regular polyhedron**   a polyhedron in which all the faces are congruent polygons, and the faces all have the same angle between them.

**rhombus**   a parallelogram with two adjacent sides equal.

**right angle**   an angle with degree measure of exactly 90°.

**right prism**   a prism in which the lateral edges are perpendicular to the plane of the base.

**ring**   the area between two concentric circles.

**rotation**   the figure resulting from two consecutive reflections through intersecting lines.

**scalene**   a triangle with all sides having different lengths.

**secant**   a line that cuts through a circle at two points.

**sector**   that portion of a circle cut off by two radii and an arc.

**segment**   that portion of a circle cut off by a chord and an arc.

**semicircle**   an arc of a circle cut off by a diameter.

**semiperimeter**   the measure of half the perimeter.

**similarity**   a relationship that preserves ratios of distances.

**skew lines**   lines in different planes which do not intersect.

**slant height**   the altitude of a triangular face of a pyramid, or any lateral element of a right circular cone.

**square**   a quadrilateral that is both a rectangle and a rhombus.

**straight angle**   an angle with rays that form a straight line; the measure of the angle is exactly 180°.

**supplementary angles**   two angles which can form a straight angle; two angles whose degree measure adds to 180°.

**syllogism**   a string of conditional statements used to create a new statement.

**tangent**   a line which intersects a circle at one and only one point.

**tetrahedron**   a polyhedron with four faces.

**theorem**   a conditional statement that is proven.

**transformation**   a one-to-one correspondence between two sets of points.

**translation**   the figure resulting from two consecutive reflections through parallel lines.

**transversal**   a line that cuts across two or more other lines.

**trapezoid**   a quadrilateral with one, and only one, pair of sides parallel.

**trisect**   to divide into three equal parts.

**Venn diagram**   circles used to represent sets; also called Euler circles.

**vertex**   the common point that is the intersection of the two rays of an angle.

**vertical angles**   the opposite pair of angles formed when two lines intersect.

**zone**   the surface area of a sphere cut off by two parallel planes.

## HERON'S SEMIPERIMETER FORMULA

Given: $\triangle ABC$ with sides $a$, $b$, $c$

Proof: $A_{ABC} = \sqrt{s(s-a)(s-b)(s-c)}$

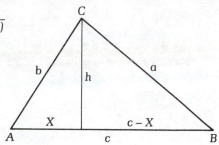

(Analysis: Use the Pythagorean Theorem in both triangles to find $X$ in terms of $a$, $b$, and $c$. Then use $X$ and $b$ to find $h$ in terms of $a$, $b$, and $c$. Lastly use $h$ and $c$ to find the area.)

1. Draw $h \perp AB$ — 1. Construction

2. $X^2 + h^2 = b^2$ — 2. Pythagorean Th.

3. $h^2 = b^2 - X^2$ — 3. Subtraction

4. $h^2 + (c - X)^2 = a^2$ — 4. Pythagorean Th.

5. $h^2 = a^2 - (c - X)^2$ — 5. Subtraction

6. $b^2 - X^2 = a^2 - (c - X)^2$ — 6. Substitution

7. $b^2 - X^2 = a^2 - c^2 + 2cX - X^2$ — 7. Substitution

8. $2cX = b^2 + c^2 - a^2$ — 8. Addition

9. $X = \dfrac{b^2 + c^2 - a^2}{2c}$ — 9. Division

Now, because $h^2 = b^2 - X^2$

10. $h^2 = (b - X)(b + X)$ — 10. Factoring (alg. subst.)

11. $h^2 = \left(b - \dfrac{b^2 + c^2 - a^2}{2c}\right)\left(b + \dfrac{b^2 + c^2 - a^2}{2c}\right)$ — 11. Substitution

12. $h^2 = \left(\dfrac{2bc - b^2 - c^2 + a^2}{2c}\right)\left(\dfrac{2bc + b^2 + c^2 - a^2}{2c}\right)$ — 12. Addition

13. $h^2 = \left(\dfrac{a^2 - (b^2 - 2bc + c^2)}{2c}\right)\left(\dfrac{(b^2 + 2bc + c^2) - a^2}{2c}\right)$ — 13. Substitution

14. $h^2 = \left(\dfrac{a^2 - (b - c)^2}{2c}\right)\left(\dfrac{(b + c)^2 - a^2}{2c}\right)$ — 14. Factoring

15. $h^2 = \dfrac{[a + (b - c)][a - (b - c)][(b + c) + a][(b + c) - a]}{4c^2}$ — 15. Factoring

16. $h^2 = \dfrac{(a + b - c)(a - b + c)(a + b + c)(-a + b + c)}{4c^2}$ — 16. Substitution

17. $\dfrac{c^2h^2}{4} = \left(\dfrac{a+b-c}{2}\right)\left(\dfrac{a-b+c}{2}\right)\left(\dfrac{a+b+c}{2}\right)\left(\dfrac{-a+b+c}{2}\right)$    17. Mult. and division

Now let $s = \dfrac{a+b+c}{2}$, the semiperimeter

18. $s - a = \dfrac{a+b+c}{2} - \dfrac{2a}{2} = \dfrac{-a+b+c}{2}$    18. Subtraction

19. $s - b = \dfrac{a+b+c}{2} - \dfrac{2b}{2} = \dfrac{a-b+c}{2}$    19. Subtraction

20. $s - c = \dfrac{a+b+c}{2} - \dfrac{2c}{2} = \dfrac{a+b-c}{2}$    20. Subtraction

21. $\dfrac{c^2h^2}{4} = s(s-a)(s-b)(s-c)$    21. Substitution (into step 17)

22. $\tfrac{1}{2}ch = \sqrt{s(s-a)(s-b)(s-c)}$    22. Substitution (= roots)

23. $A_{ABC} = \sqrt{s(s-a)(s-b)(s-c)}$    23. Subst. (area formula)

*United Feature Industries.*

## CONVERSION FACTORS—
## COMMON METRIC STANDARDS

| | | |
|---|---|---|
| 1 centimeter | = | .393701 inch |
| | = | .032808 foot |
| 1 meter | = | 39.3701 inches |
| | = | 3.28084 feet |
| | = | 1.09361 yards |
| 1 kilometer | = | .621371 mile |
| 1 inch | = | 2.54 centimeters |
| 1 foot | = | .3048 meter |
| 1 yard | = | .9144 meter |
| 1 mile | = | 1.609344 kilometers |
| 1 cu. cm | = | .061024 cu. inch |
| 1 cu. meter | = | 35.31467 cu. feet |
| 1 cu. inch | = | 16.38706 cu. cm |
| 1 cu. foot | = | .028317 cu. meter |
| | = | 7.48052 gallons |
| | = | 28.31685 liters |

| | | |
|---|---|---|
| 1 sq. cm | = | .155000 sq. inches |
| 1 sq. meter | = | 10.76391 sq. feet |
| 1 sq. inch | = | 645.16 sq. millimeters |
| 1 sq. foot | = | .092903 sq. meters |
| 1 liter | = | 1.056688 quarts |
| | = | .0353147 cu. feet |
| 1 ounce | = | 29.57373 cu. centimeters |

1 centimeter—the width of your smallest fingernail

1 meter—the length from your fingers to your nose with your head turned away

1 liter—a little more than a quart

5 grams—the weight of a nickel (5¢)

1 kilogram—the weight of a quart of milk

50 kilograms—the weight of a small woman

100 kilograms—the weight of a large man

# ANSWERS TO SELECTED PROBLEMS

Included here are the answers to the odd-numbered problems. The answers to the even-numbered problems, and to all the starred problems, are available in an instructor's manual.

Many of the answers given here are not unique—especially the proofs —there are many ways to prove some of the exercises.

For the sake of saving space, the proofs have many abbreviations, but are still sufficient to show the method used. Your instructor may expect more complete statements than are given here.

## CHAPTER 1

*Exercise Set 1*

**1.** 3.67 l    **3.** 8.8 bags    **5.** 3.9 m    **7.** Both cans hold the same amount.
**9.** 35    **11.**                     **13.** Yes. One method:

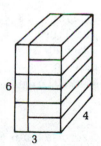

**15.** Yes    **17.** No. A box can be packed with the harmonic brick $a \times ab \times abc$ if and only if the box has dimensions $ap \times abq \times abcr$ for some natural numbers $p, q, r$ (i.e., if the box is a "multiple" of the brick).

*Exercise Set 2*

**1.** Valid    **3.** Not valid    **5.** Valid    **7.** Not valid. Does the theater consider 14 a child?    **9.** Valid    **11.** Not valid    **13.** Not valid    **15.** Valid
**17.** Not valid    **19.** Not valid    **21.** Not valid    **23.** Not valid
**25.** Valid    **27.** Expert    **29.** Change of meaning
**31.** Change of meaning    **33.** Loaded words    **35.** Loaded words
**37.** Diversion    **39.** Bandwagon    **41.** Expert

*Exercise Set 3*

**1.** a) If the alternate interior angles are equal, the two lines are parallel. b) If two lines are not parallel, the alternate interior angles are not equal. c) If the alternate interior angles are not equal, the two lines are not parallel. **3.** a) If the car does not break down, then it has had proper care. b) If the car has not had proper care, then it will break down. c) If the car breaks down, then it did not have proper care. **5.** a) If you have swine flu, then you have an urge to roll in the mud. b) If you do not have an urge to roll in the mud, then you do not have swine flu. c) If you do not have swine flu, then you will not have an urge to roll in the mud. **7.** Inverse **9.** Converse **11.** Contrapositive **13.** If it is a square, then it is a quadrilateral. **15.** If you are an elephant, then you are small. **17.** If I break my shoelace, then I will not get married. **19.** If we financially support our schools, then we will have a higher standard of living. **21.** If I wash my car, then it will be dirtier than it is now. **23.** If I go swimming, I will drown.

## CHAPTER 2

*Exercise Set 4*

**1.** Subtraction, division **3.** Substitution **5.** Halves of equals are equal. **7.** Quantities equal to the same or equal quantities are equal. (Substitution could be used twice.) **9.** Doubles of equals are equal **11.** Supplements of equals are equal **13.** Addition, division **15.** Subtraction **17.** Division **19.** Substitution **21.** Doubles of equals are equal **23.** Addition

*Exercise Set 5*

**1.** True **3.** True **5.** False **7.** False **9.** True **11.** True **13.** True **15.** True **17.** False **19.** True **21.** 29°50′46″ **23.** 60°9′14″ **25.** 150°9′14″ **27.** Through a given point only one line can be drawn parallel to a given line. **29.** 1 rad = 57.29577951° and 1° = .017453293 rad

*Exercise Set 6*

Shown here is the finished construction. Your work should also show all the construction lines. The figures are proportionally correct.

**1.**

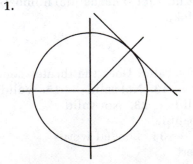

**3.** 60°

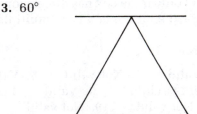

**5.** Bisect a 90° angle twice.

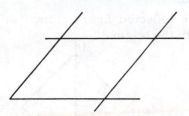

**9.** Construct parallel lines through a point on each ray of the angle.

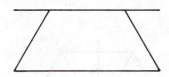

**13.** Construct parallel lines, swing equal arcs for the legs.

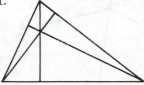

**17.** Bisect it, then bisect the two sections.

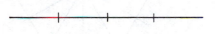

**21.** They should pass through a single point.

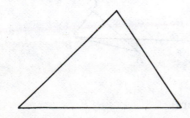

**25.** Measure the first side, swing arcs to locate the third vertex.

**7.** They should pass through a single point.

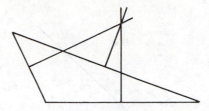

**11.** They bisect each other.

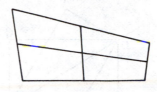

**15.** Copy ∢A twice and ∢B once.

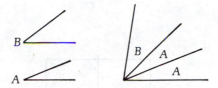

**19.** Construct perpendiculars at each end, locate the corners with equal arcs.

**23.**

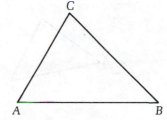

Exercise Set 7

**1.**

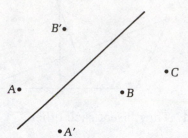
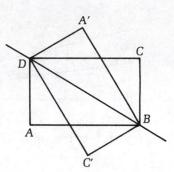

**3.**

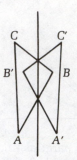

**5.**

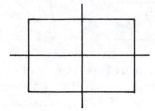

**7.** The reflected figure coincides with the original.

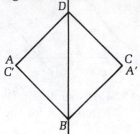

**9.**

**11.**

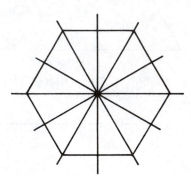

**13.**

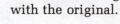

**15.**

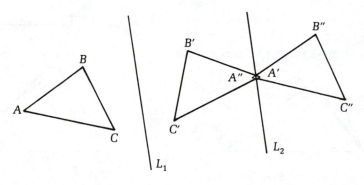

**17.** TOT, MOM, TAT, WOW (and others)

**19.**

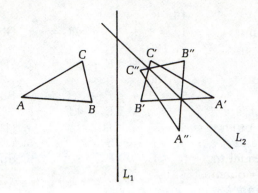

**21.**

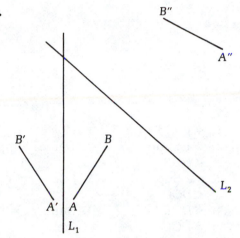

**23.** These are the digits from 1 to 9 with their reflections.

▽ 88 ꟼꟼ

**25.** Vertical: 2, 3, 4, 5, 6, 7, 8, 9, 10, A
Horizontal: 2, 4, 10

*Exercise Set 8*

**1.**

| 1. $AB$ is st. line | 1. Given |
| 2. $\angle 1$ Supp. $\angle 3$ | 2. Def. of supp. |
| 3. $\angle 2$ Supp. $\angle 4$ | 3. (2) |
| 4. $\angle 3 = \angle 4$ | 4. Given |
| 5. $\angle 1 = \angle 2$ | 5. Supp. of $=\angle$s are $=$ |

**3.**

| 1. $\angle 1 = \angle 3$ | 1. Given |
| 2. $\angle 1 = \angle 2$ | 2. Vert. angles |
| 3. $\angle 3 = \angle 4$ | 3. (2) |
| 4. $\angle 2 = \angle 4$ | 4. Subst. |

**5.**

| 1. $\angle 1$ Comp. $\angle 2$ | 1. Given |
| 2. $\angle 1 = \angle 3$ | 2. (1) |
| 3. $\angle 2 = \angle 4$ | 3. Vert. angles |
| 4. $\angle 3$ Comp. $\angle 4$ | 4. Subst. |

**7.**

| 1. $\angle BAD = \angle CAE$ | 1. Given |
| 2. $\angle CAD = \angle CAD$ | 2. Reflexive |
| 3. $\angle BAC = \angle DAE$ | 3. Subtr. |

**9.**

| 1. $\angle ABC = \angle DEF$ | 1. Given |
| 2. Both are bisected | 2. (1) |
| 3. $\angle 1 = \angle 2$ | 3. Halves of $=$s are $=$ |

**11.**

| 1. $\angle DOX + \angle XOG = \angle DOG$ | 1. Given |
| 2. $\angle DOG > \angle DOX$ | 2. Whole is $>$ any part |

**13.**

| | |
|---|---|
| 1. $\angle 1 = \angle 2$ | 1. Given |
| 2. $\angle 2 = \angle 3$ | 2. (1) |
| 3. $\angle 3 = \angle 4$ | 3. (1) |
| 4. $\angle 1 = \angle 4$ | 4. Quant. = same or = quant. are = each other. ($\angle 1$ and $\angle 4$ are equal to the = quant. $\angle 2$ and $\angle 3$) |

**15.**

| | |
|---|---|
| 1. $\angle A + \angle B + \angle C = 180$ | 1. Given |
| 2. $\angle D + \angle E + \angle F = 180$ | 2. (1) |
| 3. $\angle A + \angle B + \angle C = \angle D + \angle E + \angle F$ | 3. Subst. |
| 4. $\angle A = \angle D$ | 4. (1) |
| 5. $\angle B = \angle E$ | 5. (1) |
| 6. $\angle C = \angle F$ | 6. Subtr. |

# CHAPTER 3

*Exercise Set 9*

**1.**

| | |
|---|---|
| 1. $BD$ bisects $\angle B$ | 1. Given |
| 2. $\angle 1 = \angle 2$ | 2. Def. of bis. |
| 3. $AB = BC$ | 3. (1) |
| 4. $BD = BD$ | 4. Reflexive |
| 5. $\triangle ABD \cong \triangle DBC$ | 5. SAS |

**3.**

| | |
|---|---|
| 1. $\angle 1 = \angle 2$ | 1. Given |
| 2. $\angle 1$ Supp. $\angle DOB$ | 2. Def. of Supp. |
| 3. $\angle 2$ Supp. $\angle DOC$ | 3. (2) |
| 4. $\angle DOB = \angle DOC$ | 4. Supp. of = $\angle$s are = |
| 5. $OB = OC$ | 5. (1) |
| 6. $OD = OD$ | 6. Reflexive |
| 7. $\triangle DOB \cong \triangle DOC$ | 7. SAS |
| 8. $\angle B = \angle C$ | 8. CPCTE |

**5.**

| | |
|---|---|
| 1. $\angle 1 = \angle 2$ | 1. Given |
| 2. $\angle 1$ Supp. $\angle 3$ | 2. Def. of Supp. |
| 3. $\angle 2$ Supp. $\angle 4$ | 3. (2) |
| 4. $\angle 3 = \angle 4$ | 4. Supp. of = $\angle$s |
| 5. $AO = OD$ | 5. (1) |
| 6. $\angle 5 = \angle 6$ | 6. Vert. angles |
| 7. $\triangle ABO \cong \triangle CDO$ | 7. ASA |
| 8. $AB = CD$ | 8. CPCTE |

**7.**

| | |
|---|---|
| 1. $AD \perp$ bis. of $BC$ | 1. Given |
| 2. $BD = CD$ | 2. Def. of bis. |
| 3. $\angle BDA = \angle CDA$ | 3. $\perp$s form = adjacent $\angle$s |
| 4. $AD = AD$ | 4. Reflexive |
| 5. $\triangle ADB \cong \triangle ADC$ | 5. Rt. $\triangle$s with = legs |
| 6. $AB = AC$ | 6. CPCTE |

**9.**

| | |
|---|---|
| 1. $DF \perp$ bis. of $AB$ | 1. Given |
| 2. $\triangle ADF \cong \triangle BDF$ | 2. See #7 for method |
| 3. $AF = BF$ | 3. CPCTE |
| 4. $EF \perp$ bis. of $BC$ | 4. (1) |
| 5. $\triangle BEF \cong \triangle CEF$ | 5. (2) |
| 6. $BF = CF$ | 6. CPCTE |
| 7. $AF = CF$ | 7. Subst. |

**11.**

| | |
|---|---|
| 1. $\angle 1 = \angle 2$ | 1. Given |
| 2. $AC = BD$ | 2. (1) |
| 3. $DC = DC$ | 3. Reflexive |
| 4. $\triangle ACD \cong \triangle BDC$ | 4. SAS |

**13.**

1. ∡ABC = ∡ACB    1. Given
2. BD bis. ∡ABC    2. (1)
3. CE bis. ∡ACB    3. (1)
4. ∡DBC = ∡ECB    4. Halves of =
5. BC = BC    5. Reflexive
6. △DBC ≅ △ECB    6. ASA
7. BD = CE    7. CPCTE

**15.** 38

---

*Exercise Set 10*

**1.**

1. AB = AC    1. Given
2. ∡1 = ∡2    2. ∡s opp. = sides
3. ∡3 Supp. ∡1    3. Def. of Supp.
4. ∡4 Supp. ∡2    4. Def. of Supp.
5. ∡3 = ∡4    5. Supp. of =s are =

**3.**

1. ∡CAB = ∡CBA    1. Given
2. ∡1 = ∡2    2. (1)
3. ∡3 = ∡4    3. Subtr.
4. AP = BP    4. Sides opp. equal ∡s

**5.**

1. Draw BD    1. Construction
2. BD = BD    2. Reflexive
3. AB = CD    3. Given
4. AD = BC    4. (3)
5. △ABD ≅ △CBD    5. SSS
6. ∡A = ∡C    6. CPCTE

**7.**

1. AB = BC    1. Given
2. ∡BAC = ∡BCA    2. ∡s opp. = sides
3. AD = CD    3. (1)
4. ∡DAC = ∡DCA    4. (2)
5. ∡BAD = ∡BCD    5. Addition

**9.**

1. AB = BC    1. Given
2. ∡BAC = ∡BCA    2. ∡s opp. = sides
3. AD bis. ∡BAC    3. (1)
4. CD bis. ∡BCA    4. (1)
5. ∡DAC = ∡DCA    5. ½s of =s
6. AD = CD    6. Sides opp. = ∡s
7. △ACD is isos.    7. Def. of isos.

**11.**

Given: △ABC is isos. with AC = BC
Prove: ∡ACD = ∡BCD

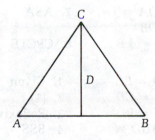

1. AC = BC    1. Given
2. ∡A = ∡B    2. ∡s opp. = sides
3. CD is a med.    3. (1)
4. AD = BD    4. Def. of median
5. △CAD ≅ △CBD    5. SAS
6. ∡ACD = ∡BCD    6. CPCTE

**1.**

| | |
|---|---|
| 1. Draw $BD$ | 1. Construction |
| 2. $AD = BC$ | 2. Given |
| 3. $AB = CD$ | 3. (2) |
| 4. $BD = BD$ | 4. Reflexive |
| 5. $\triangle ABD \cong$ $\triangle CBD$ | 5. SSS |
| 6. $\angle A = \angle C$ | 6. CPCTE |

**3.**

| | |
|---|---|
| 1. $AC = BC$ | 1. Given |
| 2. $\angle A = \angle B$ | 2. Sides opp. = angles |

(Note that the information regarding midpoints was unnecessary.)

**5.**

| | |
|---|---|
| 1. $AB = CD$ | 1. Given |
| 2. $AD = BC$ | 2. (1) |
| 3. $BD = BD$ | 3. Reflexive |
| 4. $\triangle ABD \cong$ $\triangle CBD$ | 4. SSS |
| 5. $\angle ABD =$ $\angle CDB$ | 5. CPCTE |
| 6. $PQ$ bis. $DB$ at $O$ | 6. (1) |
| 7. $DO = OB$ | 7. Def. of bis. |
| 8. $\angle DOP =$ $\angle QOB$ | 8. Vert. angles |
| 9. $\triangle PDO \cong$ $\triangle QBO$ | 9. ASA |
| 10. $PO = QO$ | 10. CPCTE |

**7.**

| | |
|---|---|
| 1. $AC = A'C$ | 1. Given |
| 2. $BA = BA'$ | 2. (1) |
| 3. $BC = BC$ | 3. Reflexive |
| 4. $\triangle ABC \cong$ $\triangle A'BC$ | 4. SSS |
| 5. $\angle 1 = \angle 2$ | 5. CPCTE |
| 6. $BC$ bis. $\angle ACA'$ | 6. Def. of bis. |
| 7. $BC \perp$ bis. of $AA'$ | 7. Bis. of vertex $\angle$ of isos. $\triangle$ is $\perp$ bis. of base |

**9.**

| | |
|---|---|
| 1. $DO = OC$ | 1. Given |
| 2. $EO = OF$ | 2. (1) |
| 3. $\angle DOE =$ $\angle COF$ | 3. Vert. $\angle$s |
| 4. $\triangle DOE \cong \triangle COF$ | 4. SAS |
| 5. $\angle D = \angle C$ | 5. CPCTE |
| 6. $\angle DOA =$ $\angle COB$ | 6. (3) |
| 7. $\triangle DOA \cong$ $\triangle COB$ | 7. ASA |
| 8. $\angle A = \angle B$ | 8. CPCTE |

**11.**

| | |
|---|---|
| 1. $AD = BC$ | 1. Given |
| 2. $AB = CD$ | 2. (1) |
| 3. $BD = BD$ | 3. Reflexive |
| 4. $\triangle ABD \cong$ $\triangle BCD$ | 4. SSS |
| 5. $\angle ADB =$ $\angle CBD$ | 5. CPCTE |
| 6. $DE = BF$ | 6. (1) |
| 7. $\triangle ADE \cong$ $\triangle CBF$ | 7. SAS |
| 8. $\angle 1 = \angle 2$ | 8. CPCTE |

**13.**

| | |
|---|---|
| 1. $AD = BC$ | 1. Given |
| 2. $AB = CD$ | 2. (1) |
| 3. $BD = BD$ | 3. Reflexive |
| 4. $\triangle ABD \cong$ $\triangle CBD$ | 4. SSS |
| 5. $\angle ABD =$ $\angle CDB$ | 5. CPCTE |
| 6. $DE = BF$ | 6. (1) |
| 7. $EF = EF$ | 7. (3) |
| 8. $DF = BE$ | 8. Addition |
| 9. $\triangle CDF \cong$ $\triangle ABE$ | 9. SAS |
| 10. $\angle DCF =$ $\angle BAE$ | 10. CPCTE |

**15.**

| | |
|---|---|
| 1. Draw $DM$ and $CM$ | 1. Construction |
| 2. $M$ is midpt. of $AB$ | 2. Given |
| 3. $AM = BM$ | 3. Def. of mid-point |
| 4. $AD = BC$ | 4. (2) |
| 5. $\angle A = \angle B$ | 5. (2) |
| 6. $\triangle DAM \cong$ $\triangle CBM$ | 6. SAS |
| 7. $DM = CM$ | 7. CPCTE |
| 8. $M$ is equidist. from $C$ and $D$ | 8. Def. of equi-dist. |

**17.**

| | |
|---|---|
| 1. Draw $PR$ | 1. Construct. |
| 2. $PQ = RS$ | 2. Given |
| 3. $QR = PS$ | 3. (2) |
| 4. $PR = PR$ | 4. Reflexive |
| 5. $\triangle PQR \cong \triangle PRS$ | 5. SSS |
| 6. $\angle Q = \angle S$ | 6. CPCTE |

**19.**

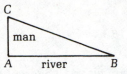

| | |
|---|---|
| 1. $AC = DF$ | 1. Height is constant |
| 2. $\angle ACB = \angle DFE$ | 2. Visor angle stays the same |
| 3. $\angle CAB = \angle FDE$ | 3. He stands erect |
| 4. $\triangle CAB \cong \triangle DEF$ | 4. ASA |
| 5. $AB = DE$ | 5. CPCTE |

*Exercise Set 12*

**1.**

| | |
|---|---|
| 1. $\angle 1 = \angle 2$ | 1. Given |
| 2. $\angle 3 = \angle 1$ | 2. Vert. angles |
| 3. $\angle 2 = \angle 3$ | 3. Subst. |
| 4. $AB \parallel CD$ | 4. Corr. $\angle$s = |

**3.**

| | |
|---|---|
| 1. $AB = CD$ | 1. Given |
| 2. $AB \parallel CD$ | 2. (1) |
| 3. $\angle 1 = \angle 2$ | 3. Alt. int. $\angle$s |
| 4. $AC = AC$ | 4. Reflexive |
| 5. $\triangle ACD \cong \triangle ACB$ | 5. SAS |

**5.**

| | |
|---|---|
| 1. $AC \parallel DE$ | 1. Given |
| 2. $\angle CAB = \angle EDB$ | 2. Corr. angles |
| 3. $BC \parallel EF$ | 3. (1) |
| 4. $\angle EFD = \angle CBA$ | 4. (2) |
| 5. $AB = DF$ | 5. (1) |
| 6. $\triangle ABC \cong \triangle DEF$ | 6. ASA |

**7.** 80°    **9.** 75°    **11.** 150°

**13.** 30°    **15.** 150°    **17.** 30°

**19.** Given: Scalene triangle with $CD$ bis. $\angle C$
      Prove: $CD$ not perp. to $AB$

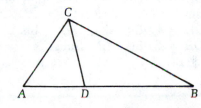

| | |
|---|---|
| 1. Assume $CD$ is perp. to $AB$ | 1. One of two possibilities |
| 2. The angles at $D$ are both rt. angles | 2. Perp. form right angles |
| 3. $CD = CD$ | 3. Reflexive |
| 4. $\angle ACD = \angle BCD$ | 4. Given |
| 5. so $\triangle ACD \cong \triangle BCD$ | 5. ASA |
| 6. and $AC = BC$ | 6. CPCTE |
| 7. But this contradicts a scalene triangle; therefore, it is *not* the case that $CD$ is perpendicular to $AB$ | 7. Given |

**21.**

| | |
|---|---|
| 1. $AC = BC$ | 1. Given |
| 2. $\angle 1 = \angle B$ | 2. $\angle$s opp. = sides |
| 3. $\angle 1 = \angle 2$ | 3. (1) |
| 4. $\angle B = \angle 2$ | 4. Subst. |
| 5. $AD \parallel BC$ | 5. = alt. int. $\angle$s |

**23.**

| | |
|---|---|
| 1. $PR = QS$ | 1. Given |
| 2. $RT = QT$ | 2. (1) |
| 3. $PT = ST$ | 3. Subtr. |

(The parallel information was not needed.)

**CHAPTER 4**

*Exercise Set 13*

**1.**

| | |
|---|---|
| 1. AE & BD are alt. | 1. Given |
| 2. ∡AEC is a rt. ∡ | 2. Def. of alt. |
| 3. ∡BDC is a rt. ∡ | 3. (2) |
| 4. ∡AEC = ∡BDC | 4. Subst. |
| 5. ∡C = ∡C | 5. Reflexive |
| 6. ∡1 = ∡2 | 6. If 2 ∡s =, the 3rd are = |

**3.**

| | |
|---|---|
| 1. ∡1 + ∡3 = 180 | 1. St. angle |
| 2. ∡2 + ∡4 = 180 | 2. (1) |
| 3. ∡1 + ∡2 + ∡3 + ∡4 = 360 | 3. Addition |
| 4. ∡1 + ∡2 = 270 | 4. Given |
| 5. ∡3 + ∡4 = 90 | 5. Subtr. |
| 6. ∡3 + ∡4 + ∡C = 180 | 6. Sum of △ |
| 7. ∡C = 90 | 7. (5) |
| 8. △ABC is a rt. △ | 8. Def. of rt. △ |

**5.** $52°10'43''$   **7.** $20°, 40°, 120°$   **9.** $115°$   **11.** $∡A + ∡B + ∡C = 180$, $9X + 3X - 6 + 11X + 2 = 180$, $X = 8$, $∡C = 11X + 2 = 90°$ so △ABC is a rt. △

**13.**

| | |
|---|---|
| 1. ∡A + ∡B + ∡C = 180 | 1. Sum of the angles of a triangle |
| 2. △ABC is equilateral | 2. Given |
| 3. ∡A = ∡B = ∡C | 3. An equilateral triangle is equiangular |
| 4. 3∡A = 180 | 4. Subst. (in step 1) |
| 5. ∡A = 60 | 5. Division |
| 6. ∡A = ∡B = ∡C = 60 | 6. Subst. (in step 3) |

**15.** $∡A = 33°, ∡B = 62°, ∡C = 85°$   **17.** $∡1 = 9°$

**19.**

| | |
|---|---|
| 1. AB = AD | 1. Given |
| 2. ∡ABD = ∡ADB | 2. ∡s opp. = sides |
| 3. BD bis. ∡ABC | 3. (1) |
| 4. ∡ABD = ∡DBC | 4. Def. of bis. |
| 5. ∡ADB = ∡DBC | 5. Subst. |
| 6. AD ∥ BC | 6. = alt. int. ∡s |

**21.**

Given: ∡B and ∡E are rt. ∡s,
    $BC = EF, ∡A = ∡D$
Prove: △ABC ≅ △DEF

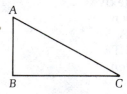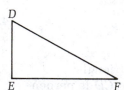

| | |
|---|---|
| 1. ∡B & ∡E are rt. ∡s | 1. Given |
| 2. ∡B = ∡E | 2. All rt. ∡s are = |
| 3. ∡A = ∡D | 3. (1) |
| 4. ∡C = ∡F | 4. If 2 ∡s are =, the 3rd are = |
| 5. BC = EF | 5. (1) |
| 6. △ABC ≅ △DEF | 6. ASA |

**23.**
Given: Alt. $AD$ = Alt. $BE$
Prove: $\triangle ABC$ is isosceles

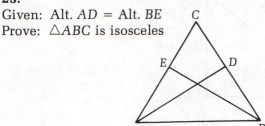

| | |
|---|---|
| 1. $AD$ and $BE$ alt. | 1. Given |
| 2. $AD \perp BC$; $BE \perp AC$ | 2. Def. of alt. |
| 3. $\angle ADC = \angle BEC$ | 3. Rt. $\angle$s are = |
| 4. $\angle C = \angle C$ | 4. Reflexive |
| 5. $AD = BE$ | 5. (1) |
| 6. $\angle CAD = \angle CBE$ | 6. 2 $\angle$s =, 3rd are = |
| 7. $\triangle CAD \cong \triangle CBE$ | 7. ASA |
| 8. $AC = BC$ | 8. CPCTE |
| 9. $\triangle ABC$ is isos. | 9. Def. of isosceles |

## Exercise Set 14

**1.** 15 **3.** 20 **5.** 75 **7.** 1001 = 1001 **9.** X = 2.5 **11.** Yes
**13.** No **15.** Addition **17.** Inversion **19.** $42\frac{6}{7}$ by 60 cm
**21.** Al: 6'$\frac{1}{2}$'', Ben: 5'6'', Chuck: 4'11$\frac{1}{2}$'' **23.** $1\frac{1}{8}$ c sugar, $\frac{3}{8}$ c butter, $1\frac{1}{2}$ eggs,
$\frac{3}{8}$ c milk, $\frac{3}{8}$ c cocoa, $1\frac{1}{2}$ c flour, $\frac{3}{4}$ c water, $1\frac{1}{2}$ t soda

## Exercise Set 15

**1.** $7\frac{1}{2}$ **3.** 2:1 **5.** 4.8 **7.** 9 **9.** 4.8 **11.** $AB = 13$, $BC = 31\frac{1}{5}$
**13.** $PQ = 250$ cm, $QR = 225$ cm, $RS = 375$ cm, $ST = 150$ cm, $PU = 300$ cm,
$UV = 270$ cm, $VW = 450$ cm, $WX = 180$ cm **15.** Parallel lines divide transversals proportionally, and if the studs are equally spaced, the transversal will
be divided into equal parts also.

**17.**

| | |
|---|---|
| 1. $AD = 5$ | 1. Given |
| 2. $CD = 10$ | 2. (1) |
| 3. $\dfrac{AD}{CD} = \dfrac{1}{2}$ | 3. Division |
| 4. $CD = 10$ | 4. (1) |
| 5. $BD = 20$ | 5. (1) |
| 6. $\dfrac{CD}{BD} = \dfrac{1}{2}$ | 6. (3) |
| 7. $\dfrac{AD}{CD} = \dfrac{CD}{BD}$ | 7. Subst. |
| 8. $CD \perp AB$ | 8. (1) |
| 9. $\angle CDA = \angle CDB$ | 9. Rt. $\angle$s are = |
| 10. $\triangle ACD \sim \triangle BCD$ | 10. $\angle = \angle$ & incl. sides prop. |

**19.**

| | |
|---|---|
| 1. Both $\triangle$s are rt. $\triangle$s | 1. Given |
| 2. The rt. $\angle$s are = | 2. Rt. $\angle$ = |
| 3. A pair of acute $\angle$s are = | 3. (1) |
| 4. The $\triangle$s are $\sim$ | 4. AA Sim |

**21.** Yes **23.** 27.4 m

**25.**

| | |
|---|---|
| 1. $CD$ bisects $\angle ACB$ | 1. Given |
| 2. $\angle ACD = \angle BCD$ | 2. Def. of bisect |
| 3. $\dfrac{AC}{CD} = \dfrac{CE}{BC}$ | 3. (1) |
| 4. $\triangle ACD \sim \triangle BCE$ | 4. SAS Similarity |
| 5. $\angle A = \angle E$ | 5. Corr. angles of similar triangles |

**27.**

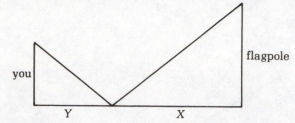

You and the flagpole are upright, that is, form right angles.

The angles at the mirror are =. (angle of incidence = angle of reflection)

The triangles are similar. AA Sim.

So knowing your own height, $h$, measure $X$ and $Y$ and solve the proportion:

$$\frac{\text{Flagpole}}{X} = \frac{h}{Y}$$

*Exercise Set 16*

**1.** $2\sqrt{14}$    **3.** $4\sqrt{13}$    **5.** $2\sqrt{13}$    **7.** $7\sqrt{2}$    **9.** $6 + 6\sqrt{3}$    **11.** 9 & 16
**13.** 54 m    **15.** $9\sqrt{2}$    **17.** $\sqrt{21}$    **19.** $14\sqrt{3}$    **21.** 282 cm
**23.** 29–420–421    **25.** $4.5\sqrt{2}$    **27.** 3–4–5, 6–8–10, 7–24–25, 8–15–17,
12–16–20, 15–20–25, 20–21–29, 36–77–85    **29.** 5 units
**31.** $\sqrt{(X_1 - X_2)^2 + (Y_1 - Y_2)^2}$

## CHAPTER 5

*Exercise Set 17*

**1.** 720°    **3.** 3240°    **5.** 9    **7.** 108°    **9.** 150°    **11.** 6    **13.** 360°
**15.** 360°    **17.** 45°    **19.** 15    **21.** 5    **23.** Yes    **25.** No    **27.** No
**29.** 12    **31.** 144°

*Exercise Set 18*

**1.** False    **3.** True    **5.** True    **7.** True    **9.** False    **11.** True
**13.** True    **15.** True    **17.** True

**19.** Given: $ABCD$ is a rectangle
     Prove: $AC = BD$

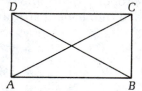

1. Draw $AC$ and $BD$      1. Construction
2. $AD = BC$      2. Opp. sides of a rect. are =
3. $\angle A$ and $\angle B$ are rt. $\angle$s      3. Def. of rectangle
4. $AB = AB$      4. Reflexive
5. $\triangle ABC \cong \triangle ABD$      5. SAS
6. $AC = BD$      6. CPCTE

**21.**
1. $ABCD$ is a $\square$      1. Given
2. $\angle A = \angle BCD$      2. Opp. $\angle$s of a $\square$
3. $AB \parallel CD$      3. Def. of $\square$
4. $\angle BCD = \angle CBE$      4. Alt. int. $\angle$s
5. $AD = BC$      5. Opp. sides of a $\square$
6. $AD = CE$      6. (1)
7. $BC = CE$      7. Subst.
8. $\angle CBE = \angle E$      8. $\angle$s opp. = sides
9. $\angle A = \angle E$      9. Subst. (steps 8, 2 & 4)

**23.**

| | |
|---|---|
| 1. ABCD is a trapezoid | 1. Given |
| 2. AB ∥ CD | 2. Def. of trap. |
| 3. ∢2 = ∢3 | 3. Alt. int. ∢s |
| 4. AD = CD | 4. (1) |
| 5. ∢1 = ∢3 | 5. ∢s opp. = sides |
| 6. ∢1 = ∢2 | 6. Subst. |
| 7. AC bis. ∢A | 7. Def. of bis. |

**25.** Given: Trapezoid ABCD with AD = BC
Prove: ∢B Supp. ∢ADC

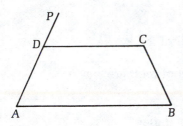

| | |
|---|---|
| 1. ABCD is isos. trapezoid | 1. Given |
| 2. Extend AD to P | 2. Construction |
| 3. ∢PDC supp. ∢ADC | 3. Form st. angle |
| 4. AB ∥ CD | 4. Def. of trapezoid |
| 5. ∢PDC = ∢A | 5. Corr. angles |
| 6. ∢A supp. ∢ADC | 6. Subst. |
| 7. ∢B = ∢A | 7. Base ∢s of isos. trapezoid |
| 8. ∢B supp. ∢ADC | 8. Subst. |

In like manner show that ∢A supp. ∢BCD

**27.** Given: E, F, G, & H are midpoints of
AD, AB, BC, & CD respectively
Prove: EFGH is a parallelogram

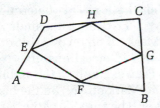

| | |
|---|---|
| 1. E, F, G, H are midpoints | 1. Given |
| 2. Draw AC | 2. Construction |
| 3. DH = HC & DE = AE | 3. Def. of midpoint |
| 4. $\dfrac{DH}{DE} = \dfrac{HC}{AE}$ | 4. Division |
| 5. EH ∥ AC | 5. A line div. 2 sides of a △ prop. is ∥ to 3rd |

In like manner show FG ∥ AC

| | |
|---|---|
| 6. EH ∥ FG | 6. Lines parallel to the same line |

In like manner show GH ∥ EF

| | |
|---|---|
| 7. EFGH is a parallelogram | 7. Its opp. sides are parallel |

**29.** Given: *ABCD* is a rectangle,
E, F, G, & H are midpoints
Prove: *EFGH* is a rhombus

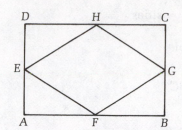

| | |
|---|---|
| 1. *EFGH* is a parallelogram | 1. By method of ex. 27 |
| 2. *ABCD* is a rectangle | 2. Given |
| 3. *AB = DC* & *AD = BC* | 3. Opp. sides of rect. are = |
| 4. E, F, G, H are mid-pts. | 4. (2) |
| 5. *AF = BF* & *AE = BG* | 5. Halves of equals are = |
| 6. $\angle A = \angle B$ | 6. Def. of rectangle |
| 7. $\triangle AFE \cong \triangle BFG$ | 7. SAS |
| 8. *EF = GF* | 8. CPCTE |
| 9. *EFGH* is a rhombus | 9. Rect. with 2 adj. sides = is a rhombus |

**35.** 91

*Exercise Set 19*

**1.** Construct a perpendicular at one end of the given side, swing arcs with side length to determine the other vertices.  **3.** Same as ex. 1 except the angle need not be a right angle.  **5.** Copy the base angle at each end of the longer base, mark off the length of the legs, join the points found for the upper base. **7.** Construct vertical angles with the given angle of intersection, bisect the diagonal lengths and mark off on the vertical angles (since the diagonals of a parallelogram bisect each other), join the points to form the parallelogram. **9.** From one end of the base swing an arc the length of the diagonal, from the other end of the base swing an arc the length of the leg. This determines an upper vertex. Do the same for the other vertex.  **11.** Construct angle *A* at *A* and *B*, extend to intersect at the vertex.  **13.** Mark *CB* off on one side of angle *C*. Swing an arc with length *AB* from *B*. The two intersections give two triangles—one acute and one obtuse (ex. 14). This is an example of the ambiguous case SSA.  **15.** Construct a 60° angle. Bisect it. Mark off "*a*" on the bisector. Construct a perpendicular at the end of "*a*". This will be one side of the hexagon. **17.** At the end of the short side construct a perpendicular. From the other end swing an arc with the length of the diagonal. This determines the longer side.

*Exercise Set 20*

**1.** $\frac{20}{121}$ or about $\frac{1}{6}$   **3.** 4 m by 6 m   **5.** 128 sq cm   **7.** $4\sqrt{2}$ cm **9.** $80\sqrt{2}$ sq cm   **11.** 10 cm   **13.** $9\sqrt{3}/4$ or $\frac{9}{4}\sqrt{3}$ sq cm   **15.** 15 sq cm **17.** 240 sq cm   **19.** $2\sqrt{14}$ sq m   **21.** 8 cm   **23.** $36\sqrt{3}$ sq cm **25.** 55 sq cm   **27.** 78 sq cm   **29.** 60 sq cm   **31.** Yes, *A* = 19,247 sq ft or about .44 acre

*Exercise Set 21*

**1.**

1. Draw $EO$      1. Construction
2. $EO = CO$      2. Radii of the same circle
3. $\angle 2 = \angle CEO$      3. $\angle$s opposite equal sides
4. $OF \parallel CE$      4. Given
5. $\angle 1 = \angle 2$      5. Corresponding angles
6. $\angle CEO = \angle 1$      6. Subst.
7. $\angle EOF = \angle CEO$      7. Alt. int. angles
8. $\angle EOF = \angle 1$      8. Subst.
9. $\overset{\frown}{EF} = \overset{\frown}{FD}$      9. = cent. angles have = arcs

**3.**

1. Draw $OP \perp AD$      1. Construction
2. $AP = PD$      2. $\perp$ from cent. bisects the chord
3. $BP = PC$      3. (2)
4. $AB = CD$      4. Subtraction

**5.** 9 cm     **7.** 1 cm     **9.** 3 cm, 7 cm, 11 cm     **11.** 8 km     **13.** $2\sqrt{3}$ units

**15.**

1. $\angle AOB = \angle COD$      1. Given
2. $\angle BOC = \angle BOC$      2. Reflexive
3. $\angle AOC = \angle BOD$      3. Addition
4. $AC = BD$      4. = cent. angles have = chords

**17.**

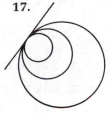

**19.**

1. $OC \perp AB$      1. Given
2. $\overset{\frown}{AD} = \overset{\frown}{BD}$      2. $\perp$ from cent. bisects the chord and its arcs
3. $\angle 1 = \angle 2$      3. = arcs have = cent. angles

**21.**

1. $OC = 5$, $CD = 12$,      1. Given
    $OD = 13$
2. $13^2 = 12^2 + 5^2$      2. Subst. (or Reflexive)
3. $(OD)^2 = (CD)^2 + (OC)^2$      3. Subst.
4. $\triangle OCD$ is a rt. $\triangle$      4. Pythagorean Thm.
5. $AB$ is a tangent      5. A line $\perp$ to a radius at its extremity is a tangent

**23.** $13(1 + \sqrt{3})$ cm     **25.** 76 cm

*Exercise Set 22*

**1.** $25°, 33°, 56°, 62\frac{1}{2}°$     **3.** $36°, 54°, 90°$     **5.** $123°, 237°$
**7.** $\angle P = 111°$, $\angle Q = 90°$, $\angle R = 69°$, $\angle S = 90°$     **9.** $125°$     **11.** $82°$
**13.** $\angle 1 = 15°$, $\angle 2 = 115°$, $\angle 3 = 35°$, $\angle 4 = 130°$

**15.**

1. $BA = BC$      1. Given
2. $\angle A = \angle C$      2. $\angle$s opp. = sides
3. $\overset{\frown}{DC} = \overset{\frown}{AE}$      3. = $\angle$s have = arcs
4. $\overset{\frown}{DE} = \overset{\frown}{DE}$      4. Reflexive
5. $\overset{\frown}{CE} = \overset{\frown}{AD}$      5. Subtraction
6. $CE = AD$      6. = arcs have = chords

**17.** One of several methods:

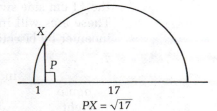

$PX = \sqrt{17}$

*Exercise Set 23*

**1.** $30\pi$ cm    **3.** 50 cm    **5.** 120°    **7.** $\dfrac{5\pi}{3}$ cm    **9.** $4\pi\sqrt{3}$ cm

**11.** $\dfrac{4\pi}{3} + \dfrac{8\sqrt{3}}{3}$ cm    **13.** $\dfrac{48\pi}{5}$ cm    **15.** $122\frac{1}{2}$°    **17.** $\dfrac{45}{2\pi}$ cm

**19.** 3 cm    **21.** The circumference is greater. $h = 3$ diameters while
$C = \pi$ diameters    **27.** $196\pi$ sq cm    **29.** $121\pi$ sq cm    **31.** $35\pi$ sq cm

**33.** $4\sqrt{3}$ cm    **35.** $\dfrac{100\pi}{3} - 25\sqrt{3}$ sq cm    **37.** Circle $= \dfrac{100}{\pi}$ sq cm,

Square $= 25$ sq cm. The circle is greater, as $\dfrac{100}{\pi} - 25 =$ about 6.83 sq cm

**39.** $88\pi$ sq cm

## CHAPTER 7

*Exercise Set 24*

**1.** 268 sq m    **3.** $10\sqrt{3}$ cm    **5.** 180 sq cm    **7.** $4\pi$ sq m    **9.** 3 cm
**11.** 11 cm    **13.** 17:30    **15.** 7 cm    **17.** 48 sq units    **19.** $20\pi$ sq units
**21.** 6200 sq cm    **23.** $152\pi$ sq cm    **25.** $288\sqrt{3}$ sq m
**27.** $5\pi(5 + \sqrt{89})$ sq cm    **29.** $625\pi$ sq cm

*Exercise Set 25*

**1.** 240 cu m    **3.** $144\sqrt{2}$ cu cm    **5.** $63\pi$ cu cm    **7.** $\dfrac{507\pi}{4}$ cu cm

**9.** $90\sqrt{3}$ cu cm    **11.** The can    **13.** 840 cu cm    **15.** $48\pi$ cu cm

**17.** 1:3    **19.** $\dfrac{32\sqrt{2}}{3}$ cu cm    **21.** $12 - 18 + 8 = 2$    **23.** $6 - 10 + 6 = 2$

**25.** $\dfrac{9\pi}{8}$ cu m    **27.** No. Angel food pan $= 96\pi$ cu. in. or about 302 cu. in.

Pyrex pan $= 256$ cu. in.    **29.** About 27.7 gal.    **31.** $16\pi$ cu units
**33.** About 317 bu.    **35.** .36 cu m
**37.** 3 green: 8,   2 green: 12,   1 green, 6,   no green: 1
**39.** $V = \dfrac{Bh}{3} = \dfrac{12 \times 12 \times 6}{3} = 288$ cu cm,   $V = \dfrac{s^3}{3} = \dfrac{12^3}{3} = 288$ cu cm

## CHAPTER 8

*Exercise Set 26*

**1.** No measurements can be exact. There are always inherent errors in the
equipment and in the observations.    **3.** Many. Lobachevskian.    **5.** Less
than 180°. Lobachevskian.    **7.** No, there cannot be "parallel" great circles.

*Exercise Set 27*

**1.** 2 & 4 cm, 3 & 6 cm, $\frac{10}{3}$ & $\frac{20}{3}$ cm    **3.** $12\sqrt{3}$ cm    **5.** From one end of
the 11 cm side swing an arc of 8 cm; from the other end swing an arc of 6 cm.
These arcs will intersect to locate the third vertex of the triangle. Locate the
incenter by bisecting any two angles.

**7.**

|         | incenter | orthocenter | centroid | circumcenter |
|---------|----------|-------------|----------|--------------|
| *acute* | in       | in          | in       | in           |
| *right* | in       | on          | in       | on           |
| *obtuse*| in       | out         | in       | out          |

**9.** $\dfrac{5\sqrt{3}}{6}$ cm, $\dfrac{5\sqrt{3}}{3}$ cm    **11.** 10 cm    **13.** $\frac{5}{2}$ units    **15.** $3(2-\sqrt{2})$ cm

**17.** $\frac{15}{2}(\sqrt{2}-1)$ cm

*Exercise Set 28*

**1.** 3 with $6+\pi$ cm, 1 with $3\pi$ cm    **3.** $\dfrac{\pi s}{2}$ units    **5.** $10\pi + 30$ cm

**7.** $\dfrac{15\pi}{2}$ cm    **9.** $9\pi$ cm    **11.** $p = 12.57$ m so 13 mowing strips will cost

$16.25    **13.** $23.47    **15.** $14\pi$ cm    **17.** $16\pi$ sq cm    **19.** $12\frac{1}{2}$ sq cm

**21.** $96\pi - 72\sqrt{3}$ sq cm    **23.** $8\sqrt{3} - 3\pi$ sq cm    **25.** $98\pi$ sq cm

**27.** 28 m    **29.** $108\pi$ sq m    **31.** $77\pi$ sq m    **33.** $300 + \dfrac{225\pi}{4}$ sq units

**35.** $80\pi$ sq units    **37.** $576 - 48\pi$ sq units    **39.** $124 + \dfrac{81\pi}{2}$ sq units

*Exercise Set 29*

**1.** From the figures: $a^2 + b^2 + 4$ triangles $= c^2 + 4$ triangles. By subtraction of four triangles $a^2 + b^2 = c^2$

**3.** In $PQRSUT$ and $OUVWXT$

$PQ = a = OU$
$QR = c = UV$
$RS = b = WX$
$SU = a = VW$
$UT = c = XT$
$TP = b = TO$ and the corr. angles can also be shown equal.

**5.** 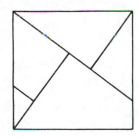      **7.**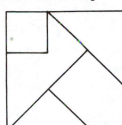

**9.** The upper edges of the trapezoids, and the hypotenuse of the triangle, all have different slopes. Though the fit is not bad, it is enough to create an error of 1 unit.

*Exercise Set 30*

**1.** Student measurement. Should be close to the golden mean.    **3.** 58.2 cm

**5.** Too short    **7.** Student's choice of Lucas sequence. Ratios approach $\phi$

**9.** 27.5 cm    **11.** $AC = 6.2$ cm, $AB = 3.8$ cm, $BC = 2.4$ cm    **13.** 7.4 cm

**15.** 97.1 cm    **17.** Student measurements. Values should be around $\phi$

**19.** $S_n = a_{n+2} - 1$ There is an equation, which you are not expected to find, for $a_n$ as a function of $n$. It is Binet's Equation:

$$a_n = \frac{\phi^n - \left(\dfrac{1}{\phi}\right)^n}{5} \qquad \text{Applying this to } S_n : S_n = \frac{\phi^{n+2} - \left(\dfrac{1}{\phi}\right)^{n+2}}{5} - 1$$

# INDEX

**413**